The Boundless Self

Space, Place, and Society
John Rennie Short, *Series Editor*

The BOUNDLESS SELF

Communication in Physical and Virtual Spaces

Paul C. Adams

SYRACUSE UNIVERSITY PRESS

First Edition 2005
05 06 07 08 09 10   6 5 4 3 2 1

The paper used in this publication meets the minimum
requirements of American National Standard for Information
Sciences—Permanence of Paper for Printed Library Materials,
ANSI Z39.48-1984.∞™

Library of Congress Cataloging-in-Publication Data
Adams, Paul C.
The boundless self : communication in physical and virtual spaces / Paul C.
Adams.— 1st ed.
p. cm. — (Space, place, and society)
Includes bibliographical references and index.
ISBN 0-8156-3056-5 (hardcover : alk. paper)
1. Communication in geography.   I. Title.   II. Series.
G70.A24 2005
910'.01'4—dc22
2004028844

Manufactured in the United States of America

To Karina and Kaitlin

Paul C. Adams is an assistant professor of geography at the University of Texas at Austin. He received his Ph.D. from the University of Wisconsin at Madison in 2003 and has also taught at Virginia Polytechnic Institute and State University, State University of New York at Albany, Texas A&M University, and McGill University. His articles on communication topics in geography have appeared in the *Annals of the Association of American Geographers, Political Geography, Geographical Review, Urban Geography,* and *Journal of Geography.*

Contents

Figures

Preface

COMMUNICATION FORGES CONNECTIONS through space that are made, rather oddly, both *by* us and *of* us. My communications are something I create, but they are also parts of myself. In fact, they are so much a part of myself that it is not clear what would remain of me if it were possible somehow to amputate my ability to communicate, not just in spoken words, but in writing and gesture, both "out" from myself to the world and "in" from the world to myself. Surely the greater part of what I call myself would be amputated by this (thankfully) theoretical procedure. Solitary confinement is the closest thing to such a procedure, and that is generally considered cruel and unusual punishment. Solitary confinement for life would be, for all but a rare monastic spirit, a fate worse than death. Yet even a monk in self-enforced solitary confinement has a book to read, which constitutes a thread of communication with worlds both real and imagined. Life without this thread becomes unimaginable.

Communication is self, and self is communication. What seems static, the self, depends in fact on the unceasing dynamic of signs, symbols, and signals coming and going. In this flow, we not only make and remake ourselves, but also make the world, or rather worlds, that we inhabit. By this I do not mean that we invent whatever worlds we care to speak about. Rather, our words and gestures and the images we make—maps, drawings, diagrams, photos, and videos—build ephemeral models of the world, touching on a few of its various elements: how it feels, how it seems to be put together, how it should be set up for shared life, and so on.

We draw and redraw the boundaries of the real, using language to specify things that are near, "Please hand me that stapler," and far, "The edge of the Milky Way Galaxy is fifty thousand light-years away." Redrawing the boundaries sometimes means shifting physical boundaries, "The distance to the edge of the galaxy is not the same in all directions," and sometimes shifting other kinds of boundaries. "Get the stapler yourself" redefines social boundaries between what I can and cannot obtain from another person, saying more about this person and myself and about the invisible space of social relations than about the stapler's position in Cartesian space.

Through communications, I actively define what it is to be me, and you actively define what it is to be you. Like notes on a piano, our self-defining actions begin to fade away as soon as they are complete, and more actions are necessary as long as the music goes on. Unlike piano notes, however, our communications *evoke* and *invoke* particular things, defining the world and giving it a kind of four-dimensional reality. Communications are therefore grounded in space and place. When I employ various languages (English, French, mathematics, photographic conventions) that are shared with a specific group of other people, I invite these people to judge my communications as valid or invalid. I implicitly associate myself with the community of persons capable of understanding the language I have used. This community has some sort of "footprint" on the earth surface, so my communication evokes a territory of communicators. The judgment my listeners pass on my communications is the result of a comparison between what I say and what they believe to be valid, true, genuine, or just. Such judgment presumes the existence of an "external" world existing outside of our communications—whether social or natural or both—and the possibility of communicating something about this actually existing world in a nonarbitrary way.

If I say "il pleut," my words reach out to any French speakers within earshot and draw them into a momentary social configuration in which we attend to the weather: Is it raining? If not, I may hear a dismissal, in

French, of my claim: "Mais non, il fait beau." If there are no Franco-phones around to disagree with me, my claim about the weather will lie dormant, as it were, neither defining the world nor implying a social collectivity. More often, my listeners and I speak the same language, use the same signs, but we disagree, which is to say we support different symbolic representations of complex bits of the world. Perhaps even so, we agree on the signal level: we all want to inspire people to act in a certain way. Communication has many layers, and the theories derived from the study of signs, as influential as they have become, are insufficient to map the space of communication in its totality.

I have called this book *The Boundless Self* because my main argument is that through communication we constantly surpass the body's physical boundaries. However, it is possible to be boundless and grounded at the same time. Let me explain. Through communication, not only do we surpass the physical body and become inextricably linked to a network of communicators, a vaguely defined community of persons who share the same signs or symbols, but also at the same time we redefine and rework our ties to the physical world. As nodes in a network of human actors that persists through time, depending materially on the earth and on its natural processes and components, we are both *extensible* and *grounded* in a complex, four-dimensional world. We are ontologically embedded in the world in a way that includes subjective, objective, and intersubjective domains of meaning.

Although I am trained as a geographer, my thesis offers a new direction for geographers rather than an outgrowth of existing work. I have cited many geographers in this book, but I do not provide a survey of the geographical studies relevant to the study of communication. Such a survey would be tangential to my main purpose, which is to develop an argument about the boundaries of the self and to show how communicative action constitutes a form of grounding in the world. In addition, I have left out such a survey because I do not write this book only or even primarily for geographers. The implications of my argument should be of interest to anyone who communicates and particularly to scholars in the social sciences and humanities who are trying to reconcile the newfound interest in place with an awareness that the link between person and place is not reducible to the location of the body, but that something deeper and more complex is entailed by the question "Where am I?"

I do specifically consider in chapter 5 the way influential geographers have written and thought of communication. I strongly challenge the assumption, based in critical theory, that to understand communication we must employ a critical approach—unpacking, deconstructing, excavating, exposing, and unsettling the operation of power in communications. To adopt this narrow and cynical vision of discourse attacks the foundation of the extensible self because the extensible self is based in and on communication; cynicism prompts disengagement by those audience members who are the most attentive and who are therefore most likely to be able to use their extensibility in creative and socially supportive ways. I argue instead for a thorough reconsideration of personal boundaries in space and time—a reconsideration based on the differentiation between signs, symbols, and signals, grounded in Jürgen Habermas's theory of communicative action. I also unapologetically advocate a comic and romantic appeal in geographical communications for the sake of promoting breadth, which in essence means opening up communications to the widest range of communicators. In keeping with this mission, I have the audacity to use the terms *trust, altruism,* and *love* when discussing characteristics of the extensible self. I promote these ideas not as a new paradigm or as a replacement for existing, critical approaches, but instead as a counterbalance to the tragic and ironic worldviews that currently prevail in "critical" geographies.

My self-proclaimed romanticism is based on a counterintuitive view of the self. If communication and self are inseparable, then surely the self is part of a network, a collection of links, not an object. We usually feel ourselves to be discrete entities, but that is not a particularly useful way of understanding the self. Delineating analytically where the self starts and where it stops, what is me as opposed to what is you, is an impossible task. Part of me lives in the languages I speak, and that part is also a part of millions of other people; when I speak, we all resonate. Every speaker can make the world live, and reciprocally the world lives for every speaker because he or she is part of various communicating groups, various *communities*. By *community,* I mean a group that shares a *common* language that not only makes communication possible, but also reaffirms the subject-object relationships that constitute a particular conception of self.

Insofar as a conception of self is largely shared with others and depends for its sustenance on their continued communicative action, the self only seems to be localized in autonomous organisms; it is a net rather than a set. If we persist in thinking of a community as a set, it is a set of junctions rather than a set of things, but a junction is nothing more than the ends of links that would not exist were they not also connected to another node. What is the node, the self, without the links? If we were able to answer this question, we would know the sound of one hand clapping.

The idea of interpenetrating selves is odd and startling at first because so often "you" and "I," "he" and "she," are locked in relations of conflict or desire or plain old confusion. How can I want someone who is not entirely separate from myself? How can I hate someone who is partly me? If people were not separate, then surely our experiences of the Other as both infuriating and enchanting and as always somewhat enigmatic should be different. But in pursuing this line of reasoning, we forget that communication itself implies sameness. When I speak and you listen, something is exchanged, but, more important, something is shared that allows my words to be recognized not only as a collection of signs, hollow and arbitrary as ciphers, but also as an experience of being in the world, *a symbol.*

The thought transcends the meaning of its constituent words. Analogously, the net of extensible people transcends its constituent selves. Both communication and community call on us to transcend narrow self-interest, to shatter known ways of doing things, to stretch the horizons of the self. Ironically, it is collectively as merged selves that we have created words and symbols that exacerbate the illusion of individual separateness in speech, writing, and existence. This constructed insularity of the expressed idea and of the expresser of ideas is perhaps founded in the instincts of the animals that we once were and to a large degree still are. As animals, we had only symbols and signals to build connections with others, except for the bonds of consumption and reproduction. The web of faunal interdependency was so tight that instincts had to preserve the self and pit it against others. Human communication introduced signs to the mix, however, opening up space and time to the imagination, but also increasing the qualitative differences between my world and your world, his world and her world. The bonds between people have become a several-million-year project, a space of endless exploration rather than a constant that pervades all of life. The question is how to respond to this constantly receding horizon.

However, the world not only offers us new horizons, but also throws up old and familiar walls whichever way we turn. If I say "il pleut" or "it's raining," the horizon of the statement is the weather itself, which is an objective condition with physical consequences. I can be wrong or deluded about the rain, about the climate, and about the material world in general. In a prolonged drought, people go hungry, and some may die. If it rains too long, people are endangered by flooding. Individual lives are fragile and constantly bounded by the body's vulnerability. Our attempts to translate the world into words push back the probability of danger, sickness, and death because words help us to *foresee* things—that is, to inhabit spaces we have not yet bodily occupied. This work to overcome horizons occurs in both space and time; we look ahead and we look beyond the known world.

The horizons of individual life are always narrower than the horizons of collective life. If our actions together, as a society or societies, exacerbate flooding or cause drought, then it is only through further communication, through better and more effective extensions of multiple selves to form wider communities, that we can bring our actions into alignment with the world we inhabit, thereby reducing its constraints on human life. In short, human biology and global ecology are two of the limiting factors on communication, contradicting whole ways of life and constructions of self if not sooner, then later, and communication responds to these boundary conditions. This is a large part of what I mean by communication's being a *grounding* force.

Social scientists reject concepts such as "reality" and "truth" because of the facility with which certain persons have turned "reality" and "truth" into tools for controlling others. But these same social scientists must eat, breathe, excrete, and die. Biological truths are the rock and soil on which the concepts of "reality" and "truth" rest. *How* do we eat, and *what* do we eat? Is the air fit to breathe? Can death be delayed by exercise? These questions begin to indicate that communication is grounded in a material reality: if I call a rock a sandwich, I may end up rather sick. If my society collectively calls polluted air "acceptable," thousands of persons may sicken and die. Through actions, we ground our communications in the real world.

Neither thought nor communication is fully grounded, however. The worlds we inhabit are always imaginary, just as they are always grounded. To be human is to move in this virtual space between the imaginary and the real. The objective world is like an asymptote we collectively and individually approach with our created, communicated world, but the journey is infinite; we are adrift between truth and fiction, and whatever we have to say to each other is a mixture of the true and the fictitious. Our drift toward the real is endless, although the direction is not arbitrary.

Just as there will never be a final explanation of the world, there will never be a final version of the self. Self is best seen as an empty boat carried along by the current of a community's encounter with the world. Selves move in this current, and the directions in which they may ultimately move are infinite.

Perhaps herein lies the reason people always try to push the world beyond its limits. Our boundless character chafes at the limitations the world imposes. The self cannot help but experience in an excessive, often morbid way its separateness, despite the connections that define it in every way. That, as Buddhists have recognized for more than two millennia, is simply part of being human. Language, as an infinite space, invites us to bend and twist the world to fit our every whim. Yet we also display an impulse to make the world finite and the self fixed and determined—a kind of rational machine or automaton—as a final solution to the mystery of being in the world. To face realistically and practically the limits placed on us by natural processes requires knowing just how thick and thin the lines are that join humanity into one vast web forming various communities and that join all of us to the material world.

In this book, I hope to get across above all the idea that communication is not just something to be shunned, feared, or, in the popular jargon, "critiqued." It is also something to be celebrated and cherished for the way it relates to our ability to trust others, our propensity to love, our occasional creativity, and the rare and mysterious impulse of altruism.

Yi-Fu Tuan, a geographer about whom I have much to say in the final chapter, has inspired my interest in these positive aspects of being in the world. He writes: "Geography has directed my attention to the world, and I have found there, for all the inanities and horrors, much that is good and beautiful" (1999, 115). His comment carries a deeper significance if we know that he has closely studied the sadism and masochism behind affectionate bonds, the biases that perpetually divide people, and people's desire to escape from reality. He continues (in the penultimate chapter of his autobiography) by weighing in on the side of optimism: "The near total neglect of the good is an egregious fault of critical social science, making even its darkest findings, paradoxically, less dark, if only because they are not contrasted with the bright lights that also make up the human picture" (1999, 115). Tuan's interest in presenting the "bright lights" is motivated by a pragmatic desire to show off the dark side of the world; he understands the aesthetic principle of figure-ground contrast.

I suspect Tuan believes that his readers must have experienced the world as a mixture of good places and bad places, encouraging conditions and depressing conditions, and that therefore the most persuasive appeal will be one that matches this ambivalent quality of experience. A less mixed rhetoric would in effect ask his readers to disregard random acts of kindness as worthless, discount genuine friendships as aberrations, and dismiss as illusory the occasional conviviality in human relations, or to see them all as vestiges of a world that is about to disappear. Perhaps he senses that this cynical type of writing itself would constitute a form of dominance much like the elegant and refined sadism he exposes in *Dominance and Affection* (1984).

Like Tuan, I believe the world should be rendered in as many shades as possible, the better to illuminate what is bright and the better to draw attention to what is dark. I suspect that people are like optical fibers that can illuminate a dark place only if they are connected to the light. This image of self as a sort of connecting fiber brings together my view of the self, my view of communication, and my view of the world.

I want to thank John Rennie Short for inviting me to contribute to his series and Syracuse University Press and their anonymous reviewer for bearing with me through the lengthy process of revision. My wife, Karina, and Rush Cohen generously provided help with the bibliography. John Pipkin raised my spirits countless times in Albany during the initial phases of the project by poking his head into my office and greeting me with a cheery "Hallo, Scholasticus!" Jonathan Smith and Dan Sui provided nourishing food for thought during my first three years in Texas. I must thank Karina and Kaitlin most of all for never losing their cool when I said for the hundredth time that I was "done" with "The Book."

The Boundless Self

I

The World of the Extensible Self

Humans are language animals.
—Yi-Fu Tuan, "Language and the Making of Place"

AS I SIT AT MY DESK WRITING, I am engaged with an audience that exists in other places and times. How easy it is to ignore this historical-geographical dimension of my work or for that matter any act of communication.

I am struggling to communicate with distant places and times: writing, deleting, replacing words and phrases as if involved in a heated discussion, although the room is silent. My audience is at once imaginary and real: at the moment of writing, it is imaginary; at the moment of reading, it is real. You are a figment of my imagination (in my time-space), but you are not *only* a figment of my imagination, having materialized somewhat like I imagined you (in your time-space). This odd link that the act of writing establishes between fantasy and reality is increasingly at the heart of human action; the self is challenged to become more effective through time and space, more capable of acting and sensing at a distance—or, following the terminology of Donald Janelle (1973), more *extensible*.

In prehistoric times, the spoken word was the overwhelmingly dominant form of communication, and it gave communication a local character. The voice quickly dissipates in open air, so a conversation is a small nucleus of activity, like the heat from a camp fire. Throughout most of historical time, communication still implied *place,* though places

1

proliferated to include all sorts of courtyards, streets, and rooms; there were lecture halls, cafés, and bars, each with its own form of speech and unspoken norms to guide that speech. In barely more than a century, however, the majority of communication situations shifted from place to nonplace contexts. Movie screens, television screens, computer screens, radios, cell phones, fax machines, and countless other technological products extend sensation and agency through space and time, pulling apart physical context (where the human body resides) and social context (where people act and sense the world around them). Symbols and signs on a two-dimensional surface—like hieroglyphs on papyrus or ink on a page—are another nonphysical place of encounter, less sexy than moving images, but comfortable and familiar after several thousand years of use. There is something almost quaint about coming together in this way, although the book shows no sign (I am grateful to say) of slipping into obsolescence.

I cannot imagine the sites of our encounter except in a vague way: you may be in a library or a living room, an office or a cafeteria. These are all likely spots in which to read a book. You are less likely to be on a boat or standing in the pouring rain, but anything is possible. I am not much concerned with this uncertainty about the site of reception, and that is normal, but I am concerned with one aspect of the geography of reception. I assume that my audience is familiar with a Western sense of space and time, a set of models for society and the individual that evolved in western Europe along with the map, the clock, mechanized transportation, and the democratic nation-state. I like to think that my words might mean something outside of that social context, but I have no way of knowing if they will be seen as brilliant, banal, ridiculous, or evil.

Of course, no two readers will be positioned in this space of the audience in the same way: male, female, young, old, white, nonwhite, wealthy, poor, healthy, medically challenged—each social category marks out its own kind of space within the larger volume of Western culture. These spaces are not just different levels of power or social status, they are different bases for drawing on elements of Western culture to construct a meaningful sense of place, space, time, and the self. You, the reader, are impossible for me, the writer, to locate precisely. As if in retaliation, I too am impossible to locate: my writing of this book spans several states, a Canadian province, and some tectonic shifts in my personal identity (new jobs, new titles, new friends, new homes).

Uncertainty is always involved in communication's linkages through space-time. One person may remember my examples, another my theories. One person may understand extensibility as a kind of action, and another may understand it as a condition that produces actions. Try as I might, I cannot produce the same idea in all members of my audience. A major reason for this (and one that spills over into countless secondary causes) is that our lifepaths—that is, our physical movements through space and our social movements through communities—have brought us "here" to this virtual encounter by many routes, and those tortuous routes have left tracks on each of us (Pred 1979, 1982, 1984).

Each of us is the sum of the social and physical contexts he or she has inhabited. These personal peregrinations that form our basis of experience have been labeled "lifepaths."[1] In the original formulation by Torsten Hägerstrand, the concept of the lifepath was a way of viewing people as processes rather than objects, as complex and constantly lengthening histories formed by the sequential occupation of specific space-time contexts. Every person inscribed a unique path through space-time, delimited by various constraints, and these meandering paths were a large component of individuality. The implication for communication is intriguing: much that you consider real must be a fiction to me, or a lie, and vice versa, because of the differences between our lifepaths and our resulting concepts of the world (Thrift 1985; Wright 1947).

Nevertheless, what I communicate to you, in your mind's eye (or ear), is not entirely different from what I hope to communicate and what I convey to others. An effective communication act transcends the divides between people. In some way it finds common ground, however

1. The earliest formulation of the lifepath concept I have found is in Torsten Hägerstrand, "What about People in Regional Science?" (1970, 10–11). Refinements and elaborations appear over the following two decades (e.g., Hägerstrand 1982, 1983).

imposing the divides of experience and perceived self-identity may be. By communicating, I not only make contact across space, but also sense a pulse that resonates with others' experiences in places and times of which I know nothing. Paths do converge at least in the virtual spaces of words and other media; our worlds touch one another despite the fact that you and I may inhabit different centuries, different strata of society, different nation-states, and in any case different states of mind.

The ancient Greek Empedocles defined God as the circle whose center is everywhere and whose circumference is nowhere; real-world communication is just the opposite—a circle tangent to everyone but lacking a center. Consequently, you and I are both *in* this virtual communication space of paper and ink, letters and words, insofar as we can find points of intersection on our respective horizons of experience and knowledge. We draw on our differences to find this grounding of sameness.

Communication and Agency

The act of communicating in writing is a form of *personal extensibility*. It connects me to many distant places and to social contexts that are also, in a metaphorical sense, far removed from my own. Of the lives of others I can know only a little, and the effect of my communication is therefore always uncertain. Writing is of course only one of many ways in which I can make a connection with distant Others. I can communicate via other media. Even more common is the simple act of consumption: when I buy an apple grown in New Zealand or strawberries grown in Mexico, I am sending a message (supporting an industry and the associated labor and environmental conditions) via the medium of money. When I vote, again I am sending various messages: one binary message regarding my preference, then (if I should be on the winning side) a more complex message sent through the representative's agency. If I am on the losing side, I can still send my opinions to the winner, but it is unlikely they will have any effect. The difference between writing and speaking, on the one hand, and voting and purchasing goods, on the other, lies in the fact that writing and speaking imply a space for response. That space is only a potential, and I may destroy it with a harsh word or by closing a book, but it is a space we might use to develop mutual understanding. You may understand me well and disagree with me, and I may understand that disagreement. Or perhaps we will find that we agree on more than one count. In either case, communication is the form of extensibility that permits us to approach one another in the virtual space of experience and the mind (Adams 1992, 1995, 1996, 1997, 1998, 1999, 2000). I cannot force you to agree with me; rapprochement in virtual space is a voluntary move. I can, however, increase the probability you will make that move by appealing to my credentials ("See, I have a Ph.D.; listen to me!"), but unless my rhetoric is convincing and the content of what I have to say matches some element or elements of your experience, you are unlikely to afford me too much authority.

The term *authority* lends insight into this twofold process of meeting in virtual space of the page, then coming together once again in the virtual space of the mind. An authority is one whose opinions are publicly certified, set front and center in the virtual space of this or that type of discussion. An authority on medieval music is not assumed to have anything special to say about cancer. Virtually any expert on any subject will greatly damage his or her reputation by giving advice on sexual technique, but even in this area society "authorizes" certain speakers, both male and female, who not only occupy television talk show sets and other physical places, but also orchestrate the creation of virtual places based on the discussion of this normally taboo subject. Authorities are people whose communications are endorsed by society (not unanimously, but across a wide range of subject positions) as long as they touch only on certain issues.

Thus, the authority, like the author, is a creature of modernity, in particular the specializing, "segmenting" aspect of modernity (Tuan 1982). As Michael Curry (1996) has pointed out, the specialized self with its particular forms of authority depends as well on specialized and particular kinds of places. The shift into what some call "postmodernity" and others call "late modernity" has done little to unsettle the power of the author. Even as the grounding in common experience on which the author stands is eroded by rapid cultural transformation, the

increasing juxtaposition of cultures owing to globalization, and the specialization of individual lifepaths, the author stands firm on this shifting
ground.

Perhaps this is because we have little else on which to anchor our
common life. Nature seems either too fragile or too malleable: a system
on the brink of collapse or just another mess that needs to be cleaned up
and turned to profitable uses. Religion gives way to the imperatives of
science, technology, pragmatism, and the market, until it serves mainly
to justify this or that group in its aggressive claims to deserve material
well-being. Government requires a kind of faith in collective life that is
growing increasingly scarce. But the authority's words cut across all divides, bearing instructions on how to avoid a heart attack, raise smart
children, get out of debt, and build a backyard gazebo. If I dismiss one
authority as a crackpot or a phony, I can always use another to back up
my claims.

Let me put these observations about my task as an author and about
authority in general into a societal perspective. To adopt the terminology of Anthony Giddens (1984), my social interactions as a citizen in a
modern society are quite often *disembedded* from physical contexts (the
contexts they would have had in the past), and they are *distanciated*—that
is, stretched out through space (see also Adams 1995; Thrift 1986). A
conversation by the water cooler is obviously a real event and a part of
my real self; in a small way, it is an act that defines my place in the world.
So, too, is a phone conversation with a colleague, client, or customer in
Toronto, Mexico City, Paris, or Tokyo.

Because sensation and agency come into play in any communication, it makes little sense to make a distinction between mediated and
unmediated communications as more or less *real* aspects of my life and
my world. Everything I draw into my sphere of activity is not only part
of my lifepath, but also a way of affecting my world. My presence in
space-time can be envisioned as a tree of connections, with my body's
position being the trunk and my communications forming links that
branch out into distant times and places. We can think of a person not as
a thing, like a brick or a houseplant, which is always one place at a given
moment, but as a tree whose "branches" come and go depending on the

task at hand. An image even more apt than the tree is the online computer with its links going in all directions through space, to old and new sites (and therefore linking to different times as well). Of course, if this second metaphor makes more sense to us now than the Paleolithic metaphor of person as vessel or the industrial-era metaphor of person as mechanism, then it is surely because regular exposure to the new technological complex has conditioned us to perceive its characteristics in ourselves (see Bolter 1984). Yet in many ways the computer metaphor is more apt because the sign-based languages we use, so different from those of animals, make us leaky vessels and poor mechanisms even while they make us excellent networkers.

Our ability to network depends on making connections in virtual space, yet the body that is at least necessary, if not equivalent, to the self, has always been in place. More precisely, it has been in places. The routinized movements of the body and the networking of the self are inextricably intertwined.

Let us say I drive to work, stay for eight hours, then drive home. My body has inscribed a line in space-time with two kinks (see figure 1.1). During this time, my agency has, in all likelihood, branched out from the office to other offices—phoning and answering the phone, e-mailing, even reading ordinary memos and letters—so that "being at work" is really more a matter of networking than being in a particular place (see figure 1.2). In this case, I might potentially handle much of my work from home, as more than 10 million Americans presently do (Hollow 2003). If I work in a factory instead of in an office, the products of my hands will spiral away into countless unseen places and affect the lives of people I will never know. Again, I am the hub of a network. Without networking, a member of Western society cannot do his or her job, and the job market reflects a premium placed on special extensibility skills. But who is in charge? Am I, or is the network?

Agency (the ability to act) is clearly a part of being human; technologies do not act of their own accord. They may function in society as "congealed ideology," in the terms of one self-proclaimed Luddite, but they do so by *changing the geography of human agency,* altering the range of physical and virtual interactions into which people are able to enter

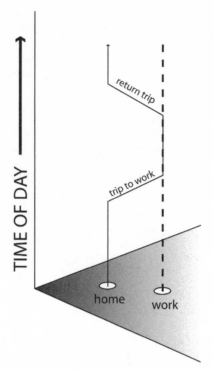

1.1. A time-space diagram with two journeys and two stations.

(Boal 1995). Therefore, it makes little sense to attribute agency (the ability to act) to communication technologies. My phone has no power until I consent to live part of my life *through* it and thereby accept the boundary conditions that it imposes on my interactions. What induces me to accept is the corresponding increase in my personal power to act across space and to affect distant places. I extend myself through the phone, and I am present in the place I call, though clearly my personal power is diminished in a certain way by calling (phoning) rather than calling (dropping in). Despite the linguistic and conceptual links between mediated and face-to-face communication, a greater level of uncertainty surrounds virtually all action at a distance. That uncertainty is something people learn to understand and manage; it does not simply negate the possibility of acting at a distance. With the right social arrangements, I can easily act through my phone: to take food out of the freezer (by call-

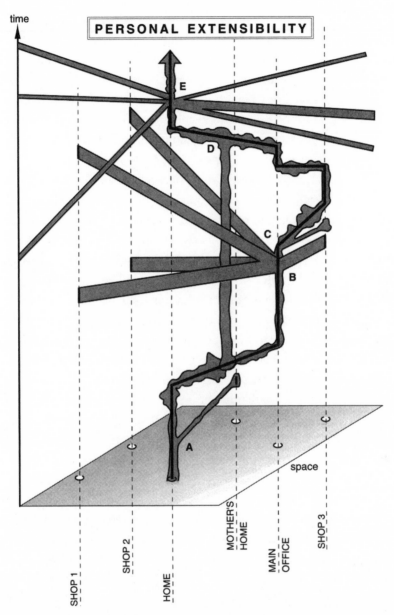

1.2. An extensibility diagram with the following simplified communication routine: *(A)* making a telephone call, *(B)* sending an e-mail message to five recipients who "open" the message at two different times, *(C)* making another telephone call, *(D)* recalling an event, and *(E)* watching the news on television. Courtesy of Blackwell Publishing Ltd.

ing my wife at home), to purchase a seat on a particular flight (by calling an airline), to organize a meeting (by calling colleagues), and so on. I can act at a distance not just because of the phone, but because it accesses spread-out parts of my activity space, where people know me and will cooperate with me or where they have access (via credit contracts) with my money. Technology transcends physical contexts for me because of a shared understanding in my social contexts and a reservoir of trust on both sides of these relationships.

Despite such shared understandings, geographical extensibility is always coupled with uncertainty, and actions become subject to increasing indeterminacy at greater distances, particularly when audiences are broader and more heterogeneous, conditions that have greatly increased with the diffusion of new media such as the Internet. From the particular geographical agent's point of view, this indeterminacy is a part of living in what Giddens calls "institutionalised risk environments" of modernity (1991, 114–37). The ability to communicate at a distance supports the coordination of activities over great distances and hence aids the control of many risks, yet the risk of miscommunication becomes progressively greater as communications spread out in space, multiply, and engage with unfamiliar social contexts. The aggregation of uncertainty can make it seem as if new media act upon society, culture, or individuals, intruding on us and disrupting our way of life or culture. In actuality, we are growing and expanding as agents in geographical and social space, and our growth is what creates the sense of disruption.

When technologies seem to act on us, we have fallen prey to a fundamental geographical misunderstanding: we have underbounded our definition of the self. We erroneously think of ourselves as points in space, and it seems to us as if this space were being invaded: wires and optical cables like the tentacles of alien beings, "wireless" radio signals like a poisonous secretion. But in fact *we are the new networks* (as well as the old places). We are cognitively, morally, and philosophically in place, and we are simultaneously spread out through space.

If my best friend asks, "Why don't you get Instant Messenger?" I know (consciously or unconsciously) that if I say "no," I risk a deterioration in my social link to that person, even a truncation of that part of myself that is defined by his or her link to me. My identity may erode ever so slightly if the friend adopts the technology and I do not. So I adopt the technology, making it part of my sphere of activity, part of myself. I am not being invaded by the technology, but instead I am invading the technology, entering the new web of social links formed in and through it. If this process seems to happen *to* me rather than being done *by* me, that is perhaps because my model of self (myself and self in general) is faulty.

Human Occupation of Space-Time

Animals, plants, and objects are always *someplace*. An animal may be in its lair, fishing by the river, climbing a tree, howling at the moon. People are not situated in space-time in quite the same way. To be a person means to live in places as we have described them, which means to inhabit virtual and physical spaces. To say "Bob" or "Lisa" is to recall a set of places where I have met with these persons, a set of places we might be able to affect in some small way, a set of places where we might get together in the future, and so on. A dog can communicate by barking or whining, but this communication does not begin to construct this kind of virtual space. Most, though not all, human communications do so.

To strip a person of all nonlocal aspects of identity would be impossible without stripping away the person as well. Even a baby has records in particular places, relatives here and there, her very own place in a crib or cradle, and so on. Of course, some social connections are lost routinely, like dead skin. Retirement and other major events may release many connections. In the United States, more than two-thirds of a person's closest social ties are severed every decade (Wellman et al. 1997). It seems that retirement would create one of the greatest disruptions of personal extensibility, but society usually affirms connections at this point rather than denying them: a person becomes a former marine, a former CEO, a professor emeritus, which does not just say something about past social ties, but also affects the sort of connections the person might forge in the future. Demotion, legal conviction, and imprisonment are more likely to sever connections, in which case time redoubles the punishment as one remembers again and again who and what was lost—amputated from activity space. Promotion, conversely, leads to more connections. Connections, of course, mean power, but we crave them for no other reason than to have them. They become a part of us and expand our sense of self.

As a professor, I carry with me certain links in time-space. I cannot treat sick people or direct traffic, but in a certain classroom I can say things and evaluate (in grades) whether my students have understood me. In effect, they enter into a legal contract to hear me or suffer the consequences. This rather odd communication situation exists in part because of what I said and did many years ago as a student in different places (the University of Colorado and the University of Wisconsin, in particular). Other professionals are in a similar situation (Pred 1979).

The self is less and less a direct outgrowth of the body or the conditions of one's family or birthplace. It is more and more a product of the intricate path a life has etched through space, the treelike structure mentioned earlier. This is not to say that the body has ceased to house the self in an often frustrating and sometimes frightening way (because the body is fragile and temporary). Nor does it mean the body is not a basis for some of the most important forms of social categorization, such as race. Significantly, legal punishment often takes the form of the captivity of the body (imprisonment) and in some states the destruction of the body (capital punishment). The body remains, of course, a primary source of pleasure in occupations as varied as eating, playing sports, hiking, dancing, swimming, and having sex. An important academic movement has therefore pressed for recognition of the body as a site of social relations (for example, see Ainley 1998; Bell and Valentine 1995; Doel and Clarke 1999; Longhurst 2001; Teather 1999). This movement's insights are most welcome because people have bodies and bodies matter in countless ways. But the potential to maintain disembodied social relationships is constantly growing. Perhaps this rediscovery of embodiment is a case of something catching our attention just as its importance begins to diminish. New media such as the Internet (and older ones such as the telephone) provide an increasing range of ways to communicate without being an embodied actor.

As Ken Hillis explains, "Human bodies . . . are an intriguing pivot for [geographical] theory, and it is difficult to imagine any geography that would matter without them" (1999, 167). Unfortunately, advocates of an "embodied" geography have been inclined to reduce the self to the body, and one claims: "Of course, in 'reality' men and women 'have' (or rather 'are') bodies" (Longhurst 2001, 13). Many current theorists equate an interest in disembodied forms of identity with sexism because, they argue, men have traditionally enjoyed the privilege of being disembodied agents, but women have been ineluctably embodied. This equation begs the argument, however, because both genders are at times stereotyped as body, and both are stereotyped as disembodied agents, albeit in different ways. Women are commonly cast as communicative, whereas men are portrayed as physical, sexual, and uncommunicative. This common stereotype suggests precisely the opposite association: male = embodied, female = disembodied. My point is that this pair of oppositions is mapped in conflicting ways, with both men and women symbolized at times by the body and at times not; to dismiss all discussions of disembodied (or embodied) agency as sexist does not serve any symbolic or strategic purpose.

We must of course acknowledge the body's reality and its inescapable limitations; its dependence on clean air, food, and water; its physical desires and pleasures; its differences from other bodies; its susceptibility to pain and discipline; its mortality; and its inescapably located character. But this acknowledgment is possible without reducing the self to the body or the body to the self.

My point can be summarized simply: more attention must be devoted to the *self as inherently contextual*. The meaning of *context* here is complicated because people are complicated beings and the spaces they inhabit take many forms. The contexts they occupy range from physical to virtual, from material to immaterial, and varieties in between. To be human, not animal, implies an awareness of the moral and ethical dilemmas of existence. As our wide-ranging minds bring us into contact with times and places we cannot see, we become aware of the consequences of our actions—if we are concerned about these consequences, as most people are, then our moral "presence" is extended through space. As always, communication guides and facilitates this moral occupation of virtual space.

The Historical Context of Extensibility

The emergence of an interregional mercantile economy in the seventeenth century, an international industrial economy in the nineteenth century, and a global information economy in the twentieth century arose from and contributed to the growth of extensibility. Geographers have referred to this progressive change as "time-space convergence" or "time-space compression," indicating a long, slow process by which distance has less of an impact on interaction owing to the speeding up of movements of people, goods, and information, which in turn depends on the creation of various fixed structures (such as railroads, phone lines, and roads) that support rapid movement of all sorts (Abler 1975a, 1975b, 1977; Harvey 1989, 1990; Janelle 1968, 1969, 1973, 1991). This evolution is not a simple forward progression, but a combination of a cyclical movement and a progression—a spiral of time-space transformation (Janelle 1969, 353). Furthermore, it is a social transformation as much as a technological transformation. New technologies are adopted at the outset of each cycle of economic expansion; as they become widespread, the economy grows, but at some point the diffusion slows along with the rate of time-space convergence, and there is a period of economic stagnation. The expectation of continued time-space convergence leads to unmet demands for access that accumulate as a kind of tension in the social system. The last phase is economic downturn, as the technologies gradually slide toward obsolescence, and the resulting crisis spurs new technological innovations (Knox and Agnew 1989, 173–75).

Extensibility is increasing if we look at it in a broad historical framework. This increase matches the multiplication of the diversity of available contexts of individual action. For many communities in the world (though certainly not all), we see a slow trajectory from a less extensible self toward a more extensible self, speaking in terms of typical or average lives. This trend of time-space convergence is closely tied to the increasing speed of new communication technologies and their growing geographical reach, as well as to the increasing volume of long-distance communication flows.

The diffusion of the printing press led to the international exchange of ideas throughout western Europe during the sixteenth century, creating a type of distanciated society that contemporaries called the "Republic of Letters" (Eisenstein 1979, 75, 138–47). Some remarkably perceptive observations on the social role of communication emerged at this time. The French statesman and minister Malesherbes stated in 1775, "What the orators of Rome and Athens were in the midst of a people assembled, . . . men of letters are in the midst of a dispersed people" (quoted in Eisenstein 1979, 132). An admirer of the new device called the telephone 114 years later described the transit of his voice: "In about one millionth of the time it takes to say Jack Robinson, it was there. It had turned a thousand curves, it had climbed up and slid down a hundred hills, and yet it came in at the finish fresh as a daisy on a dewy June morning. It was as if by a miracle the speaker had suddenly stretched his neck from New York to Boston and spoken gently into the listener's ear" (*Electrical Review* [1889], 6, quoted in Marvin 1988, 196). Such observations indicate that laypersons are aware, at least sporadically, of the process geographers describe as time-space "convergence" or "compression" (Abler 1975a; Harvey 1989; Janelle 1968, 1969). With each innovation in communication has come fleeting public awareness of the fact that previously place-based forms of community are in the process of being reconstituted over space.

We can observe the overall trend toward more rapid long-distance communication without denying that both factors of extensibility—agency and sensation—are unevenly distributed from place to place and from group to group around the globe. The growth of extensibility has advanced different degrees in different places, and the rate of change differs greatly from place to place. The wealthy are the first to benefit, which increases their competitive advantage over poorer persons whose access is delayed. The exact same dynamic pertains to relations between wealthy and poor cities, countries, and regions. Furthermore, within any given community, people benefit unequally from technological changes, depending on a wide range of factors. We have only vague ways of comparing places in regard to technological access, mainly through the use of clumsy national data on televisions per capita, telephones per capita, computers per capita, and so on. This kind of data fails to indicate differences in media access of the rich and poor, men and women, old and young, and people of various ethnic backgrounds living in any particular place (Balnaves, Donald, and Hemelryk Donald 2001).

The diffusion of communication technology is always a politicized process because of the variable constraints placed on access (some finding it easier to adopt) and the fact that such variability maps onto preexisting power distributions and social hierarchies. One critic claims starkly that "the new technologies of the virtual life are set to compound the old system of domination with fresh colonizations" (Boal 1995, 13). Yet Western history would be much more static and predictable if new media always helped the powerful segments of society solidify their social supremacy. A glance backward at the diffusion of the printing press reveals an associated loss of control by the Catholic Church and a growing threat to royalty. The sixteenth century also brought a flood of early how-to books and self-improvement manuals addressing topics as diverse as cooking, manners, medicine, art, music, surveying, architecture, engineering, and warfare (Eisenstein 1979, 242–47). The printing press may have been appropriated as a tool by those in the middle and upper segments of society, but it helped decentralize power from the aristocracy and the ecclesiastical authorities. Likewise, the Internet evolved out of a system designed to support military and scientific communication to become a tremendously broad-based system with applications ranging from international business to international terrorism (Adams and Warf 1997). The Internet is by no means radical, but neither is it reactionary, though it is appropriated by persons with reactionary and radical aims as well as by many people in between.

Navigation in Space and Virtual Space

Communication technologies are sometimes used to achieve virtual contact with dozens, thousands, even millions of others, but sometimes with only one. Some media are two way and some are one way. Some are multisensory (e.g., television extends vision and hearing), whereas some extend only a single sense (e.g., radio extends hearing, and photography extends vision). Such media variables are combined with personal "navigation" variables, such as one's skill and purpose in using a particular medium, to constitute what Arjun Appadurai (1996) calls a "mediascape"—a landscape of media access and meaning. We can rephrase this idea in terms of "virtual space," a context for interaction and sensation that does not correspond to a contiguous physical area on the earth's surface.

Everyone on the planet grows up in a mediascape. For a small percentage today, this mediascape consists mainly of speech. For most, it includes a wide array of media, including books, magazines, radio, newspapers, and television. For the wealthier members of the planetary community, the list includes computers as well. The process of growing up human now involves in most cases a restless crisscrossing of the border between physical space and virtual space. Identity is pieced together from experiences in both worlds, and, as when children carry mud and leaves into the house on the soles of their shoes, then return to the yard with toys in their hands, this boundary crossing is also a *boundary blurring*. Bits of identity picked up in virtual contexts such as television and video games adhere to the self and are carried into "real-world" contexts, where people externalize what they have previously internalized (Gerbner 2002). Influences from the "real world" conversely affect mediated experiences, determining which of those experiences will "stick" and what they will mean. Experiences in physical and virtual settings become merged and blended as they provide a basis for individual actions and personal identity.

All the while, movement between physical and virtual spaces requires a set of skills and knowledge bases that derive from both worlds. Navigation skills learned in physical spaces help children get around in virtual spaces and vice versa. For example, the general concepts of path, edge, district, node, and landmark that people have long used to structure understanding of the physical landscape can be applied in virtual spaces (Lynch 1960, 46–83). There are *landmark* publications such as

Darwin's *Origin of Species,* landmarks on the Internet such as yahoo.com, and landmarks in musical space such as the Beatles. If we read widely, our awareness of the world moves along a path of ideas from book to book, following an idea or a passion. Film buffs track the development of a particular director along a path from early to late career and note his or her landmark films. On the Internet, such paths follow hyperlinks and are forking and labyrinthine. The edges between literary genres serve as a further example of this parallel. Meanwhile, virtual spaces provide cues to the landmarks, paths, edges, and so on that one might find in physical spaces. How many Americans, presented with a photo of the Eiffel Tower would be willing to swear that it is the Eiffel Tower and that it is in Paris, without having ever visited Paris? This is just one example of the trust people place in virtual spaces as ways of knowing the real physical world.

The heuristics of navigation therefore occupy both worlds—the physical and the virtual. Likewise, the modern self—a dynamic sense of being an individual who controls his or her fate—is constructed in both physical and virtual realms. We are indelibly etched with the character of our surroundings, physical and virtual, and internalize their collective use and meaning.

Self is eminently contextual—we take our cues on who we are at any given moment from the topology of communication. Seats arranged in a circle, as in an elementary school classroom or a graduate seminar, encourage and facilitate a different sort of interaction than seats arranged differently. The circle signals to participants that every participant is an equal and that his or her contribution will be given the same weight as the contributions of others in this place. In contrast, we can think of a lecture hall, movie theater, or cathedral, where people are arranged in rows of seats fixed to the floor. The etiquette of such places is: remain silent unless one is in the place of privilege, the pulpit or lectern. Scale matters, too. A small classroom "works" differently than a large classroom. A place's scale is inflected by its density: five people in an elevator will communicate differently than five people in an audito-rium or an airplane hangar. Virtual places lack this attribute of density, but otherwise work in much the same way as physical places when it comes to structuring identity formation and interaction. An online "chat room" can be small or large (in number of participants), and whereas it is egalitarian in nature, other virtual spaces are hierarchically organized, such as spam, which is one (sender) to many (receivers) and cannot easily be resisted if one is a receiver (Adams 1998).

Personal identity is now formed within various physical places and a dizzying array of virtual spaces, reflecting their diverse and incongruent topologies. As Celeste Olalquiaga points out, it is possible to become lost in the spaces of "postmodern vicarious experience" (1992, xix). Indeed, it is doubtful if we are ever as well oriented in virtual spaces as we are in physical spaces. To act morally, however, requires a sense of place, in several senses of that phrase. What I sense as "my place" is more likely to incur a sense of responsibility than what I dismiss as having nothing to do with me. "My place" is also a place I know, and if something is not right, such as a fight or some kind of deception, I am better able to recognize it in that place and intervene. "My place" is the space in which I am willing to acknowledge my responsibility to help others, bringing that responsibility into my field of awareness while turning a blind eye to the suffering of "distant Others," people beyond the boundaries of what I define as "my place."

We are encountering a great moral and practical challenge at this point in history. In a world of expanding boundaries and growing virtualization that presents a bewildering complexity of virtual and physical spaces, and that allows virtually anyone to be included in "my place" without technical difficulty, people must work very hard to orient and reorient themselves if they are to live moral lives.

From Etiquette to Laws

Grappling with etiquette in physical space is largely a matter of understanding the norms of places and the events that take place in those places. Did I talk too loud in the restaurant? Did I sit in the wrong place at the meeting? Did I stand too close to my date at her parents' house? Did I undermine my students' respect by putting my feet up on the desk? Did I carelessly behave in my office as if I were in my living room or, worse yet, my bedroom? Such out-of-place behaviors are failures to uphold social standards, failures that are often interpreted as moral or intellectual failings (Cresswell 1996). Force lurks behind the meanings of places. Insofar as moral judgments are linked to contextually "appropriate" behavior, to be "out of place" is to run the risk of some kind of punishment or censure. Significant threats to the meaning of place, such as drinking alcohol on a city bus (private behavior in a public place) or hunting or camping on a stranger's property (public behavior in a private place) are subject to legal penalties. This system of sanctions may seem to be imposed from "above," by police and government officials, but it reflects mores and shared values relating to private and public space that the majority of the population supports. What we expect to do in a place affects what behaviors we will tolerate on the part of other people using that space and leads us to support the principle of order in the spaces and places of our lives, even if we disagree on the precise form that order should take (for an opposing view, see D. Mitchell 1995 and N. Smith 1996).

Robert Sack (1997) intriguingly points out how people seek different kinds of order in different places—an order of the mind, of the body, or of social relations. Sack carries the argument further and argues that places exert a kind of force that pulls together elements from three realms—nature, meaning, and social relations—and creates order on the planet. This argument makes sense if we always include people in our concept of place, so that "place" is not an object, but rather a way of being or existing, but I believe it is conceptually simpler and less subject to moral distortions if we understand places as simultaneously *settings for actions* and *sets for actors.*

The force a place seems to exert on us is in fact a type of indirect human agency that we impose on ourselves and others to sustain a particular set of practices and a particular repertoire of selves. Erving Goffman (1959) supports this view in his theory that as we navigate the boundaries between places, we change roles and realities, performing an elaborate, unself-conscious kind of theater. Expanding on this model, we might note that although we all try to be convincing actors in the roles we have taken on—doctor, police officer, carpenter, artist, teacher—we are obsessed with the stage sets of our lives. Plausibly presenting myself as the self I want to be is not always easy, but at least the challenges are known as long as the sets remain the same, both physically and socially. Places seem to "have" rules because our authority is so fragile; self-interest makes our own and others' misdemeanors seem naturally right or wrong. In the process of trying to conform to place-based behaviors, we make the place-behavior links seem natural to ourselves and to others. As we conform to what places seem to expect of us, we in fact *create* segmented identities to match our segmented (subdivided) world (Tuan 1982).

Modern selves are highly segmented on many axes of variation: public to private, formal to informal, unemotional to emotional, and so on. Power operates to define what are "good" and "bad" ways of occupying and using particular places; every form of sanction, from jailing to gossip, lends moral gravity to the segmentation of spaces and the associated segmentation of self (Certeau 1984; Thrift and Forbes 1983). As self-evident and natural as these forms of sociospatial segmentation seem, their purpose is to support a particular (more or less arbitrary) distribution of power in society. They do this insofar as they encourage dissymmetry between knowledge and agency and prevent certain people from achieving equal access to the spaces—physical and virtual—of public dialogues. In other words, I know of many places, but I feel capable of shaping social relations in far fewer places.

No two media of communication extend agency in exactly the same way; each supports a different kind of extensibility. To preserve one's reputation, one's friendships, and one's social connections one has to know how to behave properly in each mediated context. This issue of manners or etiquette (dubbed "netiquette" on the Internet) pushes our understanding of extensible agency toward another parallel with place and helps move us toward an awareness of the moral imperatives of extensibility. The online world is another space where I must often subordinate my will to an impersonal authority.

I agree with Jürgen Habermas (1971, 1984, 1987, 1990) that the ideal communication situation is defined by its inclusiveness, by the absence of coercion among participants, and by its rationality (which he defines in a very particular way). I also agree with Murray Low (1997) that we should find ways to promote the emergence of new kinds of nonterritorial, network-based democracy. To work toward this type of democracy requires that we expand our understanding of how communication routinely links our lives to the lives of distant others. For reasons that are not simply scientific or philosophical, then, we need to understand the spaces we inhabit—both physical and virtual—and to understand how both kinds of contexts are real in the sense that we act through them and internalize what we experience in them. Our understanding of context is a prerequisite of informed participation in civil society and is incumbent on us if we intend to act responsibly and ethically. We must recognize all of this *collectively*, not just individually, if we are to implement effective policies addressing inequality of access to media and to the public sphere.

The best way to start is with a more exhaustive look at the extensible self. What is it to be human? This is the topic of chapter 2. What personality characteristics develop in a society where there is a high value on extensibility? How do communities reflect the tensions between old and new constructions of the individual? How does communication relate to personal involvement in various communities? My discussion begins by looking at extensibility as a constantly fluctuating engagement with different scales of social integration, patterned by various routines. From here, I examine the role of extensibility in work and education, which clearly illustrates that extensibility is a form of power separate from, though often constitutive of, other sources of personal power such as wealth and knowledge. Despite these ties to power, extensibility demands personal qualities of trust, love, altruism, and creativity, which temper to a certain degree the harm an extensible individual can bring to distant lives.

Following this query into the nature of the individual, I consider the content of communication. Chapter 3 shows that signs, symbols, and signals permeate our environments and asks how we weave them together. Which elements of the awareness of place do we share with animals and which are uniquely human? What are some of the signs, signals, and symbols in the urban landscape, and what do they say about the connection between power and communication? In addressing these questions, if not thoroughly resolving them, we discover that texts such as books and places such as cathedrals have something in common. Both are complex arrangements of signs, symbols, and signals, all of which collectively add up to a single overarching symbol: in the case of a book, that symbol is the unique meaning of the book, irreducible to a smaller set of signs or symbols; in the case of a religious building, it is the belief system of the structure's users. What this commonality indicates is not only a close parallel between place and text, but also an important characteristic of sign communication: although its basic elements, words, are arbitrary associations, they are combined to form nonarbitrary associations, mappings of experience in four-dimensional space-time that reflect objective conditions of the real world. But if in theory places and texts are parallel, in actuality each is paradoxically contained within the other. At the end of chapter 3, we take a walk through the city and see how places and texts are interwoven.

Chapter 4 shifts from content to context, addressing some persistent though rather protean concerns that have resurfaced in social theory in various guises over the past five decades. What relationship between subject and object is established by a particular form of communication, and is this relationship open to reinterpretation by individuals? If we were to create an ideal communication situation, what form would it take? These questions lead us through materialist analyses of communication from Marx to Habermas and reveal the many ways social power relations affect communication content and processes. One insight this exploration provides is that the opacity of much academic writing is deliberate. The objective of intervening in societal processes is linked to arcane and sometimes incomprehensible language, betraying a profound cynicism about the communicator's power to use language to critique the existing social power relations. Habermas takes a more optimistic view, based on a number of assumptions about communicative action—the cycle of symbolization–action–symbolic revision. His view supports a "grounded" view of communication. Following hyperlinks on the Internet, we trace a path through virtual space and eventually encounter a discussion that closely approximates a Habermasian "ideal speech situation," only to discover that the dialogue is less than ideal (in the general sense of the term), indicating that his model is not quite sufficient to ensure constructive debate.

Finally, in chapter 5 I link extensibility, content, and context in a single framework. The emphasis is on developing breadth in our communications, which implies the creation of physical and virtual spaces that are open and inclusive. Minimally, this means a wide range of participants can engage in dialogue in these spaces without fear of humiliation, dismissal, or coercion by others. I argue that the democratic inclusivity of such spaces depends on including four types of breadth in our communications: social, geographical, rhetorical, and moral. Although the context of communication is important, these elements of communication content are essential if we are to contribute constructively to the creation of a global society of extensible individuals—in other words, if our discourse is to be rational and progressive. These forms of breadth together promote a kind of symmetry between our space of sensation and our space of agency, which is the basis of a moral life under the influence of globalization.

2

The Anatomy of the Extensible Self

> The dialectics of *here* and *there* has been promoted to the rank of an absolutism according to which these unfortunate adverbs of place are endowed with unsupervised powers of ontological determination.
>
> —Gaston Bachelard, *The Poetics of Space*

FOR A PERSON LIVING IN PREHISTORY, extensibility was internalized in childhood as part of the language-acquisition process, then modified slowly and incrementally throughout life. Any particular act of speech reached the ears of those persons close by, and the power to build social connections was limited by the range of the unaided voice and by the movement of the body. The linguistic abilities that made social networks possible were developed in childhood until they reached a fairly constant adult level. Extensibility was closely tied to place and was a stable component of identity because devices did not exist that could create an instantaneous link to distant people. There was no way of communicating across time except in images, such as petroglyphs, that had predetermined meanings. Communication acts were therefore intimately tied to the here and now.

As discussed in chapter 3, this does not mean communication was *about* the here and now. People could talk about the past and future and about distant places, constructing a virtual world in the mind, but the senders and receivers of any communication act were limited to the present moment and to the range of the unaided voice.

This chapter considers the history of extensibility, with its founda-

25

tions in spoken language and various forms of record keeping, and the recent proliferation of technologies facilitating distanciated interaction. It also addresses the role of education, which does not simply reflect the link between extensibility and power, but helps perpetuate it. Extensibility may not be taught as a subject in school, yet its basic skills permeate grade school curricula. These skills take obvious forms such as reading, writing, and computer skills. More elusive are the personality traits that propel people to live satisfied and meaningful lives in a society where many interactions are disembedded from physical contexts and spread out through technological networks. Trust, love, altruism, and creativity are defining features of the extensible personality, even if they seem often to be in short supply. The chapter therefore considers elements of the extensible self in historical perspective.

The Evolution of Distanciated Society

Pictograms were the earliest form of writing. They were distinguished from earlier pictures by their stereotyped character: a fixed set of symbolic images for a fixed set of meanings—a stick figure to mean "person," a circle to mean bread, and so on. Pictograms have an interesting predecessor. A system of small clay objects was employed as early as 7500 B.C.E. in the area stretching from Anatolia through the Middle East to present-day Iran and the southern shores of the Caspian Sea. The objects included disks, cones, spheres, and obtuse triangles (see figure 2.1) Denise Schmandt-Besserat, a professor of art history and Middle Eastern studies, discovered not only the purpose of these tokens, but also the fact that the early cuneiform writing used in Sumeria (Mesopotamia) corresponds closely to the shapes of these tokens in at least thirty-three instances (1978, 59). According to Schmandt-Besserat, the tokens were almost certainly used for administration and tax-accounting purposes, but their wide distribution suggests some involvement in trade or at least long-distance shipping as well. These "tokens" were originally smooth on the surface, then later incised with lines and cross-hatching to indicate distinctions of meaning. The "jar" token could be elaborated to indicate a jar of oil, for example, and a garment

2.1. Complex tokens from Susa, Iran, ca. 3300 B.C.E. *Top row, left to right:* parabola = fringed garment, triangle with five lines = ingot of metal, ovoid = jar of special oil, disk with cross = male sheep. *Bottom row, left to right:* biconoid = honey, rectangle (meaning not known), parabola = garment. Courtesy of Dr. Denise Schmandt-Besserat.

token could be elaborated to symbolize a particular kind of garment. These tangible predecessors of the written word bridged the gap between the earliest agricultural societies and the urban societies that developed out of them, just as they occupied the conceptual terrain between orality and literacy.

Furthermore, it cannot be a coincidence that the earliest urban societies developed precisely within the zone where tokens had already taken hold. But what exactly is the connection between record keeping and the radical transformation marked by the emergence of urban life? Although the development of cities is often attributed to a food surplus produced by the use of agricultural technologies, to the need to organize labor for engineering tasks such as irrigation, and to the adaptation of certain elements of the hunting and gathering society, a fundamental element of urban society is overlooked. A jump in the scale of social coordination and geographical integration demanded some sort of medium other than the spoken word, which would facilitate the organization of material life (Mumford 1961; Wittfogel 1957). Tokens were one solution to the problem, and though the use of physical objects for this purpose seems awkward and archaic to us now (because of the demands for lightness and rapid transmissibility that we associate with data-recording devices), clay tokens offered sufficient lightness and transmissibility for their time. Their users eventually encapsulated them in clay balls or strung them on strings to facilitate their management and movement. From these objects, the step to written record keeping was not far.

At some point, the record was translated from three dimensions to two dimensions, the space it has occupied ever since. Sumerian cuneiform, like Egyptian hieroglyphs and Chinese pictograms, employed a mixture of simplified pictures and arbitrary signs to stand for words or ideas, most of them originally based on a corresponding token—a circle for a disk, a triangle for a cone, and so on. The evolutionary link to the tokens is indicated by the fact that some symbols were originally created by pressing certain tokens into the flat clay tablets, but over time all symbols in this system were made by pressing a triangular stylus into the clay. The pictograms were gradually abstracted; they lost most of their pictorial characteristics and became ideograms. This evolution from pictograms to ideograms occurred in China and to a lesser degree in Egypt.

In general, an ideographic writing system such as hieroglyphs or cuneiform promotes a narrow monopoly on power because the sheer number of word signs and their lack of connection to spoken language extends the time taken to acquire literacy, which in turn maintains knowledge in the hands of the few. Furthermore, it is unlikely for such a system to be as comprehensive in terms of vocabulary size as any spoken language. Royal records and data could be preserved in this fashion, but it was not strongly supportive of the construction of personal identity or autonomy. In fact, by facilitating the growth of long-distance administration and trade, ideographic writing facilitated the formation of empire and a concurrent loss of personal power in the regions subsumed by these empires.

Yet another aspect of ideographic writing resists the spread of literacy: the treatment of writing as sacred. The term *hieroglyph* means "sacred carving," and the implication of magic power pervaded not only Egyptian hieroglyphs, but also Mesopotamian cuneiform and early Chinese ideograms, which were used for divination. When the written form of a word stands for the idea behind the whole word rather than for the sounds that make up the word and is therefore both unique and indivisible, perhaps people are more inclined to confer on that written word a kind of magical link with the signified. The arbitrariness of each sign and of signs in general is more difficult to recognize. The select group with mastery over the written language acquires a sanctified status that belies the technical character of its expertise.

A crucial challenge to this magical-aristocratic writing system came from a collection of consonant sound signs used by the Bronze Age inhabitants of Ugarit (in present-day Syria) between 4000 and 3000 B.C.E. Based on an even earlier proto-Canaanite model, but inscribed with a stylus like Babylonian cuneiform, these sound signs were the precursors of both the earliest alphabets and later alphabets, including those of classical Greece and Rome and thereafter those of Europe. The shift to widespread use of sound signs occurred among a people at the margins of the great empires of the time—Egyptian, Sumerian, and Hittite. Marginality is perhaps a key to the switch because the scribes of Ugarit apparently felt the need to communicate in four different languages and were proficient in seven different scripts. As with the Canaanites, their survival depended more on flexibility and adaptability than on rigidity and tradition. Any given writing system could hardly seem sacred when juxtaposed with six others in the Ugarit scribes' world. Furthermore, in such a multilingual world, the shift to representing words in terms of sounds would provide a mechanism of linguistic assimilation and unification because even unfamiliar words could be quickly incorporated into the writing system in this way.

By all indications, a system of sound signs—that is, an alphabet—is not a natural way for people to visualize speech. All known alphabets diffused from the linear Phoenician archetypes—Roman, Cyrillic, Hebrew, Arabic, Chinese Pinyin, Viking Futhark (runes)—although some alphabets borrowed only the concept and not the shapes. Although each alphabet in its way has been used to promote centralized religious and political power, the spread of alphabetic writing has facilitated the acquisition of literacy and thereby decentralized power from a small elite to a larger public. The transition from sacred inscriptions to a utilitarian tool for social participation was slow, and some alphabets, such as the Futhark, contain letters that are explicitly linked to various deities. When used broadly by a literate population for purposes of commerce, news dissemination, science, law, and entertainment, these mythical-magical associations tend to be diluted or wholly effaced.

An obstacle to mass literacy is scarcity of written material. Long after the diffusion of the alphabet, the majority of the population remained illiterate. It was the printing press that made literacy rates higher than 90 percent a possibility. With access to cheap printed materials written in an alphabetic language, people could relatively quickly learn to read and write and acquire reading material. Therefore, a powerful technique for extensibility was at their disposal. The implications of this set of techniques and technologies—alphabetic writing and the printing press—were radical. The family Bible was originally a foundation for intellectual autonomy; from personal interpretation came questioning of church dogma and ultimately religious reformation (Anderson 1983, 37–40; Eisenstein 1979, 303–450).

If the hold of ecclesiastical power was loosened in some places and shattered in others, the reformation of political power was not far behind. Broadsheets and printed essays fomented rebellion against the Crown in England, France, the North American colonies, and elsewhere. As late as the 1980s in Czechoslovakia, the printed word has been the primary foundation for political dissent (Havel 1991). The power of a few radicals to release their ideas from the here and now of bodily presence and to reach out quickly to the world through the medium of print means that questions can no longer be silenced, even by execution. Although Michel Foucault (1979) sees a new regime of power in the replacement of corporal punishment by imprisonment, the shift also reflects the fact that the self had slipped the bounds of the body.

Telegraphy drove a wedge between communication and transportation. Prior to the invention of the telegraph in the 1830s, someone had to move with a message in order for it to be transported, and the limits on human movement (speed, safety, cost, etc.) also limited communication. As James Carey (1989) argues, the telegraph affected transportation, investment, morals, and many other aspects of society.

During the twentieth century, extensibility evolved from a state of being to a ceaseless activity. In its new manifestation, it was a constantly fluctuating engagement with different scales of social integration, patterned by various routines. The routines linked specific technologies with particular times and places—the newspaper with breakfast with home, the radio with the car with the route to work, the television with the evening with the family and the home. Various communities in society appropriated each new medium according to their needs and in ways that extended certain aspects of preexisting social relations (Williams 1974). At the same time, each new medium created openings for creative redefinitions of the self, which in turn loosened preexisting authority structures (Meyrowitz 1985). Little can be said in general about the diffusion of communication technologies, then, because each technology has been adopted in ways that are contingent on culture and on the adopters' individual agency. The few things we can say involve the prevalence of communication technologies in daily routines, the concentration of schooling and training on learning how to use communication technologies and techniques, and the (partial) development of certain personal character traits suited to a world of extensible agents.

Communication Routines, Community, and Identity

Many people in the United States awaken to the sound of a clock radio: chat, music, and news stream into private space from beyond the walls of the home. These sounds form a link between geographical scales: the scale of the listening place (bedroom), the scale of the entire audience (up to a few thousand square miles), and the scale of the content, of what is talked or sung about (shading off from the well-known surroundings to a fuzzy and indefinite world "out there" across oceans, mountain ranges, and deserts). Of course, these geographical scales vary from individual to individual: an individual's listening place may be tucked into the ear with tiny earphones, blasted throughout a warehouse by a battered portable radio, piped through the ceiling speakers in a dentist's office, or blared to several thousand concert-goers. The scale of the content also varies: the traffic and weather report are scaled to the size of the city, whereas the hit song that follows is scaled to the size of the nation or to an international "community" of fans.

When one chooses a radio station, it is at some level an act of reaching out to the metropolitan, regional, or national scales. One can choose news from National Public Radio, which focuses on national and international affairs, or from local stations that focus on local weather, local traffic, local crime, local politics, and local special events. Radio is of course only one set of links in a complex communication network. In sharing information and comparing opinions via a wide range of media, modern communities are formed—scattered, incoherent communities, but communities nonetheless.

The quality of media-supported community is affected by the fact that mass communication is grounded in capitalism in most places. Media schedules are tailored to benefit advertisers and not audiences. For example, during "drive time" directly before and after the regular work day, commuters (who turn on the radio to escape from a traffic jam into a virtual auditory space) are likely to encounter ads or chatter instead of music. Radio puts a peculiar spin on capitalism: music is delivered free over the airwaves so as to create an audience. This audience is then sold to advertisers. In short, each audience member's extensibility is viewed as a resource that can be manipulated, bought, and sold, which shapes both the kind of community that media can support and the associated constructions of self.

Examined in aggregate, "modern social life is characterised by profound processes of the reorganization of time and space, coupled to the expansion of disembedding mechanisms—mechanisms which prise social relations free from the hold of specific locales, recombining them across wide time-space distances" (Giddens 1991, 2). Nigel Thrift (1985) has traced the beginning of this process to the distribution of "chapbooks" (cheaply bound books) and encyclopedias in England and

France during the three and a half centuries following the invention of the printing press. The impact of the emergence of literacy in tandem with the mass production of books "signified the beginning of a gradual transition from a predominantly oral to a print culture and from practical learning and reckoning to a more rationalised, systematic and distanciated view of the world" (1985, 382–83). As worldviews became more rational and systematic, the self became more mysterious and enigmatic; with so many different lives and lifestyles available to imitate, one's own life no longer appeared predetermined by family or fate. Personal biographies became a succession of choices—a "garden of forking paths," to use Borges's (1988) evocative phrase . The implications of this garden of forking paths for the constitution of personal identity are significant: people find themselves challenged and are simultaneously offered new resources to face that challenge. People must "negotiate lifestyle choices among a diversity of options" (Giddens 1984, 5). Arjun Appadurai elaborates on this point in *Modernity at Large:*

> Until recently, whatever the force of social change, a case could be made that social life was largely inertial, that traditions provided a relatively finite set of possible lives, and that fantasy and imagination were residual practices, confined to special persons or domains, restricted to special moments or places. In general, imagination and fantasy were antidotes to the finitude of social experience. In the past two decades, as the deterritorialization of persons, images, and ideas has taken on new force, this weight has imperceptibly shifted. More persons throughout the world see their lives through the prisms of the possible lives offered by mass media in all their forms. That is, fantasy is now a social practice; it enters, in a host of ways, into the fabrication of social lives for many people in many societies. (1996, 53–54)

However artificial the models of self presented by the media may be, for reasons having to do with economic exploitation above all, these models' potential is nonetheless to liberate people from the narrow confines of identity defined by tradition. The media seem to provide a transparent window onto alternative lives, and these lives not only promote consumption, but also inspire people to reconceive themselves as extensible agents, becoming more powerful, autonomous, and self-conscious. Thought and identity become impossible for the state, family, or community to determine. The self remains embedded in community—but which community? The potential answers to this question multiply as communication contexts multiply. Local community is one option out of many; commitment is divided between place-based and virtual (but still real) communities. People are always at once *citizens* and *consumers,* two somewhat contradictory models of self, and most are *workers* as well. These various aspects of the situation have profound effects on personal power.

Extensibility and Personal Power

The daily routines of agency, both in and out of the workplace, blend communications at various scales through various media to form a particular personal rhythm of involvement in various scales of social integration (Adams 1999, 2000). The workday in a developed country begins with acts of communication that quite often extend the self as a node of sensation and agency into spaces beyond the proximate environment. Someone asks her secretary for a report of the previous day's European sales. Someone else checks his e-mail. Someone else lifts a stack of letters and reports from throughout the United States and sorts it according to urgency. The high end of the service economy selects and promotes workers with skills in extensibility—the ability to read quickly, to use a computer, to develop and maintain hierarchies, to treat abstract and disembedded social situations as real. Even for "unskilled" jobs, the demands for know-how in communication are often extensive—faxing, mailing, data entry, and so on.

Notwithstanding this observation, the variation in extensible agents' time-space rhythms is remarkable: one person will make hundreds of business calls each day, whereas someone else in an equally "important" job may make only five or six calls a day. Consider the difference in extensibility between a doctor, who spends the day seeing patients in her office, and a newspaper foreign correspondent who flies to several foreign countries each month and sends home information for audiences thousands of miles away. Yet not all extensibility meets the eye. The doctor in her examining room must draw on the collected knowledge of tens of thousands of other physicians, biologists, and pharmacists in order to diagnose an illness correctly and to prescribe the best treatment. The apparently in-place communication of the examining room is, thanks to the institutionalized networks that disseminate medical knowledge to practitioners, actually a nodal point in a communication network.

Some low-paid workers, such as secretaries, must be highly extensible, whereas others, such as warehouse workers, can pass an entire day communicating only face to face. Again, this distinction begins to dissolve when we look closer: the warehouse workers sort and organize products in ways that will facilitate product movement, helping the warehouse function effectively as a node in a network of flows of goods, money, and information. They are extensible workers, albeit in an anonymous, collective way. Even at this level, some awareness of spatial interconnections is present, as workers talk about the Illinois Insurance Company order or the Franklin County contract. All workers in a distanciated society generally support distanciation in some way and benefit from skills related to more abstract or more concrete aspects of distanciation. Of course, not all workers shape or direct their own extensibility any more than all workers set their own hours or have the chance to decorate their workplaces.

Tracking Down Agency

It is somewhat artificial to focus on the individual worker in a modern economy. People pool their agency in any factory, office, or other collective work situation. This pooling implies an asymmetrical kind of sharing. The "boss" extends his or her agency out to the world through the labor of the entire office. His agency is multiplied tenfold, a hundredfold, perhaps even a thousandfold. Employees' agency is not multiplied in this way and therefore is more difficult to identify.

Let us imagine that a secretary from Office A calls a clerk at Office B to pass on information from a "higher-up," yet the information transfer typically occupies only a small part of their conversation. The clerk makes a comment about the weather or the upcoming holidays or a football game; the secretary passes on a joke she heard at a going-away party for another office worker or passes on a bit of gossip; such comments balloon into conversations, and the total phone time exceeds by several minutes the moments of actual information transfer. This communication, of course, does little to support the objectives of the "higher-up," but employees are less controllable than machines. Although the clerk and the secretary are acting for their respective bosses and their organizations, they do not simply lose consciousness and become machines; they also act as individuals. Their extensible interactions are both self-directed and externally directed.[1]

Pooled agency consists of the agency of particular persons, but separating out the contributions is impractical. Actual conversations would have to be dissected almost word by word to separate actions taken on behalf of another from actions taken in response to one's own goals and interests. More generally, we must walk a fine line between equating extensibility with economic power and stressing its collective, communal role.

What is clear is that the ability to access diverse sites and situations in particular ways leads to a good income in a technologically complex society, and income is generative of social power. But extensibility is not rewarding just because it may lead to a good income. Many movie actors and actresses continue working long after their economic security is established. The same is true of writers and politicians, musicians and pundits. Celebrities host benefit concerts simply to have an impact on the world, or so they hope. Many retired professors continue publishing and teaching. Extensibility is its own reward. Just as a villager in the preliterate world might have enjoyed social influence in his or her village, a highly extensible member of a society numbering in the hundreds of millions may simply enjoy being seen and heard by others as its own re-

1. We must not jump to conclusions about the gendering of this delegation of agency. As women increasingly take administrative positions, the men working "under" them extend their agency just as female secretaries traditionally extended the agency of their male "superiors."

ward. The fact that communication is a source of pleasure argues against reducing distanciated communication to the economic relations (such as employment and advertising) woven through and around it. The graffiti artist's impulse to "get up" in the public eye is not alien to a large proportion of humanity.

Michael Mann (1986) argues that four sources of power—political power, economic power, military power, and ideological/religious power—have structured the rise and fall of social orders throughout history. The balance between these sources of power shifts among different times and places—a point recognized in the most flexible models of power currently used by social scientists. But extensibility produces power of a special kind. Extensibility manifests itself variously as the ability to command (political power); the ability to persuade (ideological power); the ability to coerce physically, to retaliate, and to punish (military power); and the ability to harness capital flows and exploit labor (economic power). Extensibility is a more fundamental source of power than money, weapons, or persuasive ideas because it may manifest itself as any of these sources.

Let us say the world appears to demand of a person a certain amount of military power on the way to success (as in ancient Mesopotamia); therefore, he or she must necessarily marshal a supply of ideological power whenever it is necessary for crossing into virtual space (claiming divinity) and extending his or her agency. Mann's distinctions become even less clear-cut in a world where communication is often distanciated, and powerful people of all types must contend with the power of "the media." Military leaders such as U.S. secretary of defense Donald Rumsfeld or al Qaida leader Osama Bin Laden are also ideological leaders because a loss in credibility can undermine their ability to command and, at a larger scale, to influence. A head of state may be nominally a political and military leader, but in actuality he or she assumes the mantle of economic and ideological leadership. The ability to act at a distance in all possible ways is a measure of his or her effectiveness. Success in many different fields becomes a measure of a generalizable gift for extensibility—one that is capable of mutating from one form to another.

These observations apply not only in the areas of work and production, but also in the areas of leisure and consumption. Speaking of the ordinary middle-class resident of a Texas city, Miguel de Oliver and Michael Yoder claim: "Thus, an individual can, in a single evening, enjoy a Mexican meal in a New York Yankees baseball cap before driving a rural all-terrain vehicle in leather moccasins to a rap-dancing club in an Anglo part of San Antonio, and the next day rearrange the whole array of commercialized symbols" (2000, 88).

By consuming symbols, one often takes on a fantasy identity from an Other who is willfully misunderstood, such as the stereotypical Mexican depicted on the restaurant's menu: elevated and put down at the same time, both respected and trivialized. Extensibility in the form of consumption often draws a veil over the distant consequences of our actions even as it appears to reward and honor the Other. Consumption subsidizes all sorts of labor relations in distant places, which may in turn facilitate the economic exploitation and political domination of distant workers (real Mexicans harvesting avocados). Profits from international sales can be used to subsidize deceptive media and disciplinary apparatuses to subjugate workers. Yet whether one purchases Italian shoes, Chinese plastic goods, or guacamole made from Mexican avocados the link between distanciation and coercion is denied by the advertising and forgotten by the consumer. The problem suggested here is a kind of asymmetry between personal identities: the worker and the consumer, the agent and his or her sensory awareness of the world. The extensible individual affects more than he cares to know about. The extensible worker is aware of the multitude of unpleasant jobs, yet as a consumer he brackets this knowledge out of awareness and experiences the products of distant labor nostalgically or absentmindedly, as if the distant labor he depends on were itself a form of leisure. Because of these asymmetries, consumption is overwhelmingly important in the global imbalance of power. Communication systems often block signals about this situation from reaching the United States even as they treat consumption as a form of "self-expression" that marks one's position in the landscape of taste (Bourdieu 1984).

Training, Discipline, and Extensibility

In a world of extensible selves, formal education not surprisingly stresses the skills of navigation in virtual worlds. Children in elementary schools are taught to read, write, calculate, think scientifically, and use the computer. There is much they do not learn. They do not learn how to build a shelter out of sticks and leaves, identify edible roots and berries, or catch fish. They do not learn how to butcher a wild pig or chew on its hide to make leather. They do not even learn a modern urban vocation such as plumbing or car repair. They learn first of all to *represent* the world variously in portable, durable, and transmissible forms of virtual reality: numbers, letters, diagrams, graphs, and pictures. This training in virtual-reality techniques is far from indulgence in fantasy.

Our society stresses extensibility in education because the techniques of extensibility have become essential to individual *and* collective survival and success. Schools prepare people to inhabit a society where interactions are often distanciated and social situations are often disembedded; they make extensible individuals (Giddens 1984). Such training helps people adapt to and feel at home in an increasingly segmented *and* integrated world, but it also transmits a particular concept of action. Action in fact *means* something different now than it did in premodern times. A child must take action to read a book, turning the pages and scanning them with her eyes, sitting still to take in information—all of which implies bodily discipline. *Thus, a highly regimented physical action is the cost of the most lucrative types of social action.* Parents enforce this association: "Don't fidget, just read the book!"

All forms of communication require a particular physical regime: the training of the fingers for writing and typing, the training of the eye for following lines of text, the training of the ear to pick out subtleties in voice on the telephone, the training of the eyes and ears together so one can take notes in a lecture or play an instrument in a marching band. Perception of two-dimensional drawings requires a perceptual adaptation; it is a learned skill (Gregory and Gombrich 1973). Considerable perceptual adaptation is required to use media such as the architectural blue-print, radio telescope, word processor, or medical X ray. Marshall McLuhan claimed perceptively that "Media, by altering the environment, evoke in us unique ratios of sense perceptions. The extension of any one sense alters the way we think and act—the way we perceive the world" (McLuhan and Fiore 1967, 41).

Through training in techniques of extensibility, something is lost and something is gained in regard to the sensory environment and personal freedom. Such training often costs money (as well as time) and in turn makes money (and often saves time). Lewis Mumford reveals the more general cultural change associated with science and technology:

> In general, the practice of the physical sciences meant an intensifica-tion of the senses: the eye had never before been so sharp, the ear so keen, the hand so accurate. . . . But with this gain in accuracy, went a deformation of experience as a whole. The instruments of science were helpless in the realm of qualities. The qualitative was reduced to the subjective: the subjective was dismissed as unreal, and the unseen and unmeasurable non-existent. . . . As the outer world of perception grew in importance, the inner world of feeling became more and more impotent. ([1934] 1963, 49)

It would be a gross oversimplification to blame this perceptual change on capitalism alone. The change reflects a broad, multifaceted evolution in technology, social organization, political structure, and economic co-ordination. It would also be erroneous to argue that on the whole people have been disempowered by this change.

The disciplining of the body has attracted a great deal of attention since Michel Foucault darkly and strikingly described it in the 1960s, but the followers of Foucault have overlooked the self *as both active and extensible* even as they have highlighted the body as a nexus of power and knowledge. What is antithetical to this approach is the idea that the self may be given a greater range of autonomy (and hence power) through bodily discipline, an insight that depends fundamentally on the idea that the body and the self are not the same. "Docile bodies" are useful not only to existing powers, but also to the selves that reside in them (Foucault 1979). The better the training, the more the associated power, and we must wonder if bodies really are docile or if their intransigence simply takes on an extensible form.

To clarify this point, let us say the book my daughter is reading in school is about dinosaurs, in which case she is taking in information gleaned from the activities of hundreds of paleontologists working in many digs and laboratories all over the world for more than a century. She senses something of these diverse moments in the history of paleontology, and her reading extends her sensory range in an even more marvelous way to "see" something like the world that existed tens of millions of years ago with rather different seas and continents. She imagines she is a paleontologist and soon has dug a hole in the yard, or imagines she is a dinosaur and has knocked over the end table. The reader's docile body is now the virtual dinosaur's unruly body. Her unruliness may extend to the philosophical level as well if she has been raised as a "creationist" or is attending a Catholic school.

The disciplined body can free the self from the local and the familiar in a wide range of ways. To interpret training and discipline as means of perpetuating docility (through the dominating power of capital, authority, bureaucracy, or surveillance) denies the active involvement of the self in steering its own biography and in constructing both physical and distanciated interaction situations. Foucault, unlike many who use his ideas, was aware of this situation: "In reality, power in its exercise goes much further [than the state apparatus], passes through much finer channels, and is much more ambiguous, since each individual has at his disposal a certain power, and for that very reason can also act as the vehicle for transmitting a wider power" (1980, 72).

In terms of Appadurai's imagined biographies, the same disciplined movement—the manual-optical movement of the reader—lays the foundation for a career in social work, business management, environmental conservation, petroleum engineering, or law. Self-discipline, which is to say discipline over both mind and body, facilitates building the distanciated networks implied by each of these careers. If society coerces training in self-discipline, it is in part a product of consensus that this training is good for the self and the society. (As an aside, I would note that the learned immobility of studious extensibility harms the body unless it is counterbalanced by another bodily discipline such as a sport, exercise routine, yoga, or tai-chi, which are diffusing into Western society in response to this need.)

Distanciated action depends not only on technologies, but also on *techniques*—systematic ways of doing things. Morse code is a technique that made the technology of the telegraph key useful. For the technology of the computer, we have (for now) the technique of typing (borrowed from the typewriter) and manipulation of the mouse or pointer, which constitute entirely new skills in historic terms. Each technique is an interface between the individual and his or her distanciated social contexts; where there is interaction with media, techniques of extensibility are always necessary. The sum of available techniques is the mode of action of an extensible self actively (though not always freely) engaging with the nonlocal world.

Techniques are not just physical routines such as typing. Each technique arises one way or another from a single intellectual "movement" toward abstraction. An alphabet is an abstraction from speech sounds; its representation in the form of digital signals or Morse code is a further abstraction; the use of digital signals to portray a three-dimensional object such as a building in the two-dimensional space of a computer screen is yet another abstraction. In summary, distanciated communication depends on technologies and techniques, and the techniques include a wide range of disciplined physical routines that radiate out from the single key intellectual move of abstraction. Personal power arises directly out of the internalized discipline that makes all this possible, and this power expresses itself in physical and virtual spaces. Now we turn to the question of unequal power.

Extensibility and Economic Inequality

Every child receives a different level of education. In some countries, the discrepancies are relatively minor, whereas in others, such as the United States, the discrepancies are major. The poor in the United

States go to schools with few resources and consequently have fewer opportunities than the wealthy to familiarize themselves with the techniques and technologies of extensibility. This lack leads to a lower level of motivation to impart such skills to others. For the poor, education moves at a slower pace owing to lack of equipment, less-skilled teachers, and lower student motivation (itself a reflection quite often of low levels of parental extensibility). The end of education comes sooner for most of these students, either through high school attrition or the rising cost of a college education.[2] Greater economic need means they must work sooner in life and often for longer hours, therefore bypassing opportunities to acquire the most valued skills of extensibility. Those with a bit more money can afford to attend a technical college, where the training is targeted to "practical" skills, which means skills for those who use extensible technologies but do not direct their use (those whom Castells refers to as the "networked," as distinct from the "networkers" [1996, 244]). Persons with the most family money in the United States can afford to attend four-year colleges, where they learn the skills to direct others in coordinated, institutional forms of extensibility. They become engineers, accountants, business executives, lawyers, consultants, professors, and media professionals—jobs that reward the investment in techniques of extensibility with high salaries and in many cases with high demands for self-discipline. Of course, the uneven distribution of educational opportunities does not dictate that its beneficiaries will act in ways to perpetuate the underlying logic of disciplinary regimes, social hierarchy, and extensibility. Some networkers will act in opposition to the very inequities that helped them arrive where they are in the social hierarchy. In general, however, extensibility begets extensibility just as capital begets capital, and the two circuits are closely interlinked.

We might hope to simplify this picture by advancing the hypothesis that extensibility is simply a kind of capital and that therefore it merits no special approach or theory. The uneven distribution of wealth per-

2. It is, after all, at the college level where the most prized forms of extensibility are acquired, and one interpretation of a college education is as a means of transforming a finite amount of capital into a potentially unlimited amount of extensibility (quite a sensible investment under both economistic and broader rationales).

haps is simply reflected and magnified in differential access to communication opportunities and skills. Alternatively, and more intriguingly, we might pursue the idea that capital and extensibility are mutually reinforcing but fundamentally different forms of social power. Jürgen Habermas views capital as a kind of communication—*communication emptied of qualitative content and deprived of its progressive political role*. On this account, communication includes the flow of capital, but involves many other kinds of flows.

A key confirmation of Habermas's view is that not all who are trained in communication are effective communicators. Communication skill goes beyond formal training. "Born communicators" can achieve phenomenal levels of extensibility, whatever their start-up capital and credentials. The part of extensibility that cannot be taught is the part relating to individual awareness of others and of their situations. The "genius" we attribute to great speakers, artists, and political leaders includes the mysterious ability to transcend the particular *cognitively*, to speak, write, paint, or make film in a way that is somehow universal despite the limitation of the self to a single lifepath. This recognition of the art of communication above all militates against the conflation of capital and extensibility. It also creates problems for any methodology that seeks to consider people as interchangeable workers or capitalists. The only methodology capable of capturing the nature of communication as both shared resource and discourse, on the one hand, and personal expression, on the other, is a hybrid of the general and the particular, analysis and description. Therefore, Hägerstrand argues that "on the continuum between biography and aggregate statistics, there is a twilight zone to be explored" (1970, 9).

Emotional Training

I have not yet spoken of emotions, but surely emotional training as much as physical and perceptual training is required for a person to live in a distanciated society. Some of this training takes place at school, but much of it occurs outside of school in the home, in public places, and in places of consumption. I would identify trust, love, altruism, and creativity as four fundamental emotional traits entrained at some level in every extensible self. These traits are learned in physical and virtual contexts, but not in a way that can be predicted or engineered. There are, no doubt, more traits, but these traits I believe are the crucial ones. Of course, other traits such as avarice, greed, ambition, and aggression benefit people in a capitalist society, but they are not necessarily useful for extensibility. They in fact lead to a focus on accumulation and the building of defensive walls around the self, which work against the trend of extensibility.

The first trait, trust, is necessary if one is to adjust psychologically to life in a distanciated society. To some degree, all relatively well-adjusted members of a distanciated society extend their trust to countless unseen and unknown Others. The second, love, is frustratingly elusive, often quite temporary in its effects, and profoundly wasteful of productive energy, yet it is valued very highly by a distanciated society for reasons we try to analyze rather than treating it as "natural." The third, altruism, may be a special expression of either love or trust, but its distinctive character justifies special attention. The fourth, creativity, is an attribute of self and an orientation relative to space and time that is richly rewarded in a highly distanciated society. I consider each trait in turn.

Trust

Trust's purpose in modern society is to resolve some of the existential tensions of the highly extensible individual. Social theorists have identified trust as one of the most important character traits developed by people in large and complex societies because it is a major factor in determining the overall success of a society in providing for its members' needs (Dasgupta and Serageldin 2000; Rorty 1998). As such, trust appears to be a vital ingredient in economic development, but the meaning of trust for the individual goes beyond economic utility. Here, we must turn to Habermas, who reveals that trust is central to the formation of individual identity. His theory, as summarized by Thomas McCarthy, is that

Members of our species become individuals in and through being so-
cialized into networks of reciprocal social relations, so that personal
identity is from the start interwoven with relations of mutual recognition. This interdependence brings with it a reciprocal vulnerability
that calls for *guarantees of mutual consideration* to preserve both the in-
tegrity of individuals and the web of interpersonal relations in which
they form and maintain their identities. (1990, x, emphasis added)

Here we see trust in the form of "mutual consideration" as a general condition of human life. But life in a distanciated society makes
unique demands on trust. It demands daily reliance on a host of social
systems that must be maintained by a multitude of unknown and often
distant others. Consider flying in an airplane: the pilot and copilot, air-
traffic controllers, ground crew, and maintenance workers must perform in a perfectly reliable and competent fashion to avoid delay or
disaster. Even the passengers have their part to play; they are trusted to
ride quietly, follow directions, and refrain from interfering with the pi-
lots and flight attendants. Part of the shock that resulted from the attacks of September 2001 was the sudden collapse of the trust that
passengers had felt unconsciously for each other; the exceptional quality
of this period, however, indicates that "normality" in Western society
implies mutual trust.

The opportunity to buy a bag of groceries that are relatively free of
toxins, contaminants, and spoilage is also dependent on the reliability
and competence of multitudes of unknown others. Driving on the freeway, using an elevator, investing in stocks, undergoing surgery, learning
about world events through the news, using a microwave oven or power
tool—all these and many other activities in a distanciated society require
trust in others, some close at hand and known, many others distant and
unknown, in various factories, farms, laboratories, and offices. Their reliability is often a matter of life and death. The attempt to classify cultures on the basis of their level of trust—with more developed cultures
having higher trust levels—has been critiqued, but it is nonetheless clear
that individuals need a special kind of trust to live contentedly in high
modernity where distanciated interrelations are common (Krishna

2000, 75). *Terrorism* is normally a vague and heavily politicized term, but the term generally indicates a highly disruptive kind of attack on a distanciated society—one that undermines trust.

Because self-serving behavior, anomie, hostility, and other manifestations of distrust lie behind many of the dysfunctions of large societies, it is easy to overlook the pervasiveness of trust (Simmel 1961; Wirth 1938). But it is the general prevalence of trust that makes the presence of any of these forms of nontrust so devastating.

Trust is established *through* communication, but also constantly manifested *in* communication, particularly in everyday, routine communication that seems devoid of relevant content. On the road, drivers must trust each other to signal intentions in some manner. When watching television news, viewers must trust that what they are seeing is not a wholesale fabrication. Ordinary hellos and good-byes, like the use of turn signals on automobiles, build a basis of trust that is almost atmospheric in its pervasiveness. Anthony Giddens reminds us that trust is "the main emotional support of a defensive carapace or *protective cocoon* which all normal individuals carry around with them as the means whereby they are able to get on with the affairs of day-to-day life" (1991, 40, emphasis in original). Tuan (1988) demonstrates the various ways that moral values appear in the modern urban landscape, showing that although bureaucracies may be annoyingly impersonal, they are often for that very reason trustable. Perhaps only because trust is ubiquitous, it is easy to disregard or deny.

One hardly needs to point out that in a capitalist system trust is unevenly distributed, but it would be simplistic to claim that the poor can afford less trust than the rich or, conversely, that the poor suffer from an excess of trust, whereas the rich wisely curtail their trust. Both deterministic views miss the mark·because trust, like other attitudes, is not wholly determined by economic structure, but is a precondition and product of certain kinds of individual agency. The backstabbing behavior of the rich and famous and the loyalty of "common folk" to each other are two clichés. Research on social capital puts the lie to these myths, however; the diamond trade, a vocation of the very rich, is based on huge amounts of trust, and the alleviation of poverty is often frustrated by low levels of trust among the impoverished (Coleman 2000; Krishna 2000). In all probability, it is mistaken to claim that the wealthy have more or less trust in comparison to other groups.

What about trust between the wealthy and the poor, the powerful and the weak? Scholars normally dismiss the notion of such solidarity, yet Richard Sennett (1980) not only finds trust between persons with unequal shares of power, but argues that these links up and down the social hierarchy perpetuate the autonomy of both the weak *and* the powerful. His argument might be dismissed as a conservative justification of social inequality, but its validity is descriptive (objective) rather than normative (intersubjective). People do in fact seek out individuals to whom they surrender a portion of their autonomy—political leaders, military leaders, ideologues, athletes, sex symbols, and cultural icons—and a condition of this hero-worshiping behavior is the implicit trust placed by the weak in a chosen personification of power. We may try to mitigate the harm produced by this human tendency through democratic and economic institutions, but the desire to trust those who are more powerful is directed, not created, by these institutions.

If schools in highly distanciated societies focus on the training of self-discipline and skills (perhaps in that order of emphasis), they also prepare future members of society to be trustworthy and trusting. This inculcation of trust occurs in countless ways. From the kindergarten fire station field trip to the enforcement of university honor codes, students are constructed as trusting and trustworthy selves (and punished if they fail to meet this standard). These lessons generally incorporate a sense of the wider world as both a locus of trustable authorities and a kind of jury passing judgment on one's trustworthiness. To trust is an option; to be trustable is not.

The calculating reciprocity that motivates exchanges of goods and favors in a relatively closed community no longer dominates in Western society, though it still occupies offices and other workplaces (Tuan 1986b). This narrow "I'll scratch your back if you scratch mine" reciprocity demands a certain kind of trust, but in a large and complex society another kind of trust is required. People in distanciated societies find themselves giving favors that cannot be reciprocated except by a third party or parties, which turns the "circle of reciprocation" into an open, branching structure (Tuan 1986b, 15). The emotional stance that takes such acts for granted and contributes to the circulation of favors among strangers is an especially impersonal and inclusive kind of trust.

Trust bridges a gap in knowledge caused by distance or boundaries, and this gap is geographical as well as social and postulates that the intentions and abilities of others who lie outside of one's firsthand knowledge are harmless or beneficial. Even as one is being exploited by a society (say, by its industrial employment system), one also can place trust in many aspects of that society: its ability to protect from military attack, to supply heating oil, to create job opportunities, to establish and redefine the status quo, and so on. Trust is also internalized to form a personality trait; trustworthiness is an expression of one's own willingness to be depended on as a member of a large society. And the two (trust and trustworthiness) are closely tied: those who spread doubt (such as social critics) are often branded as suspect and untrustworthy characters. Although social critics serve an important function, we can perhaps appreciate the effort to silence these critics because it is motivated in part by the very personal awareness that any erosion of one's own trust for others and for society in general can have costs in regard to one's ability to inspire trust and to maintain supportive connections.

In a related vein, Deleuze and Guattari discuss the "social machine," which they consider a set of distributed flows that imitate and even seem to replace the organs of the human body (1983, 139–271). The social machine is an integration of desire, a great flow that decodes everything specific and turns it into money. As such, it is a kind of limit (everything is dissolved into the flow), and therefore the human identities it encourages are also at limits, the limits of predefined sanity: either neurosis or schizophrenia. To relate fully to the social machine one must avoid the neurosis of isolation and merge cognitively with one's environment in a condition the authors misleadingly label "schizophrenia." This worldview entails losing track of the boundaries between self and nonself, human and nonhuman, even living and inanimate. Their argument is overdrawn—ordinary reality in high modernity is certainly far from clinical schizophrenia—but we can glean from it a key insight: an empowered self in a distanciated society is comfortable with the blurring of boundaries between the self and the Other in geographical, social, and symbolic space. The trusting self who does not seek to resolve such boundary questions is not the isolated and tragic individual, which Deleuze and Guattari call "Oedipus" (and critically identify as the goal of Freudian psychotherapy), but rather the fully engaged and *desiring* agent.

The vocabulary in this model, in particular the call for "schizoanalysis," cannot take us very far because schizophrenics are impaired in their ability to empathize with others, whereas trust depends on empathy. As Habermas argues, "only to the extent to which the interpreter also grasps the *reasons* why the author's utterances seemed rational to the author himself does he understand what the author meant" (1990, 30). Empathy is the basis of communication, and empathy implies more than simply exchanging information or schizophrenically losing track of the boundaries between self and not-self; it implies striving (albeit with imperfect success) to put oneself in another's *place*. I must trust that what I hear/read means something familiar and friendly, even if framed in slightly unfamiliar terms, accents, and rhetoric. Trust is the form of belief that denies the gap between lifepaths, so trust is essential to a society where each life is a unique collage of contacts and experiences in a diverse array of physical and virtual spaces.

Love

Love is obviously related to trust, as a kind of extreme case. But whereas trust simply accommodates the mind to feel comfortable with action at a distance, love actively desires what is distant or different or both. Although love implies far more than romantic love, it is perhaps most distinct in this form. Romantic love is the enemy of distance and would seem in theory to imply a certain tension with distanciation and extensibility. However, romantic love occupies center stage in distanciated societies.

Resolving this seeming paradox requires a bit of careful thought, yet love is a topic that has attracted virtually no attention from geographers (with the notable exception of Yi-Fu Tuan).[3] My first tentative steps to understand love as a geographical subject can provide only a glimpse of what many geographers engaged in dialogue about this subject might find. I see love—both "brotherly" and "romantic"—as a prerequisite for the development of the extensible self and therefore essential to the maintenance of large-scale social integration. These forms of love have been ascendant since the Renaissance and have slowly displaced filial piety, patriarchy, and direct ownership of another person (as in slavery and feudal control) as the primary interpersonal attachments. Because the self no longer has a fixed *place* (geographical location or social place), it attaches to another person as its "home"—a shifting anchor in a world full of flux (Fromm 1956). The romantic love object is a form of personal grounding that evolves in parallel with the well-known shifts in space and place that have accompanied modernization, such as the increase in mobility, the convergence and compression of space, and the view of place as location. Attachments to place and attachments to persons are complexly interconnected and cannot be isolated one from the other, yet the partial detachment of self-identity from place, as indicated by the growth in mobility and in cosmopolitan attitudes, is in some sense contingent on redefining self around a different point of reference—a significant other. The task is to better understand the precise nature of that "significance."

Love is a shared symbol that circulates in an evolving social context. The symbol of love is easily internalized and thereby becomes a basis for the development of one's sense of self through the mutual exchange of experiences with a loved one, many of them formed through extensibility. As love has been given higher priority in societal processes, the concept of duty in interpersonal relations has been limited and circumscribed, along with the rigidity of social roles. The individual, now divided internally among a wide range of roles and situations (perhaps

3. Tuan treats love of place in *Topophilia* (1974), writes of personal love in his plaintive autobiography *Who Am I?* (1999), and discusses both sexual and platonic love in his idiosyncratic and provocative *Dear Colleague* letters (2002, 171–74).

we should call him or her a "dividual"), senses the freedom to become whatever he or she dreams of being (invariably sensing more freedom than is actually present, but thereby forgetting social limitations for a time), and he or she begins to experiment or to play with forming and dissolving bonds with others. In this process, the self is faced with many questions emerging out of the confrontation with another extensible self—questions about values, beliefs, aesthetic preferences, chosen career, whether to keep or break off relations with various other persons, and so on (Gergen 1991). Relationships based on duty and place rather than on voluntary selection and outreach have not disappeared, but they are increasingly confined to narrow areas of social relations such as child rearing.

Where social ties are left so much to the individual's discretion, a condition of normlessness or "anomie" is present: if I decide what rules to follow, these rules are not really rules, but merely preferences (Durkheim 1897). But this anomie is by no means a constant state; it comes and goes throughout life, with peaks that are socially recognized. Persons moving through one peak of anomie are labeled "teenagers," and their normlessness is widely tolerated, whereas persons moving through another phase of normlessness are said to be in a "midlife crisis," and their adolescent behavior is subject to formal and informal judgments no less than other more stable adults. Throughout the more "normal" phases of life, the extensible individual may pass by many opportunities to forge new social ties, voluntarily (though perhaps largely subconsciously) limiting new ties to exclude strong love attachments. The extensible individual's biography passes through ups and downs defined not only by significant relationships, but also by the openness to deep or shallow relationships at certain points in time. The individual sends out tendrils of self, merging his or her identity with others and incorporating their extensibility patterns into his or her own sphere of self, expanding at some times more than others, in self-timed but rather predictable phases.

The picture I have painted here is of an individual extending love (in a broad sense) not through a sense of obligation, but as a free choice.

Not surprisingly, many elements of this view have been critiqued both theoretically and morally. A few decades ago observers raised an alarm about narcissism and expressed concern that self-centered love relations threatened both the foundations of society and the autonomy of the individual. Psychologist Erich Fromm (1956) not only called loving an art, but also sternly noted that this art was generally neglected insofar as most people treated their loved ones as commodities, things to be possessed rather than foundations for self-discovery. The historian Christopher Lasch (1978) launched an attack on economic liberalism under the rubric "the culture of narcissism" and traced the origins of this self-centered hedonism to the liberal bureaucratization of society and the creation of the welfare state. Although from opposite ends of the political spectrum, both Fromm and Lasch perceived a breakdown of fundamental human relationships in Western society and a failure to love masked by a misguided form of love. Richard Sennett (1978) proclaimed the "fall of public man," lamenting the cult of personality that had replaced more formal role playing in public life and criticizing the intense interpersonal relationships that he believed had replaced public life. For Sennett, the fundamental problem was a lack of balance between private and public worlds caused by an overemphasis on love at the expense of cool and detached modes of engagement. These arguments could be synthesized to mount a critique of love within modernization: what we see is collapse of attachment and involvement to a narrow sphere—from the cosmos to the hearth, from civil society to the narcissistic and possessive microcosm of family and friends. A further implication is that although media constitute a public sphere, they do so by translating public relationships into symbols and images derived from interpersonal relationships. Finally, our love relationships themselves are tailored to fit a consumer society in that they reflect the possessive and acquisitive stance of the self constituted primarily as a consumer. These authors would not approve, perhaps, of being cobbled together like this, yet they all sound an alarm about the failure of love within a corrupted society.

Fromm's work is deeply flawed, yet it contains some of the most important insights about love. He sees romantic love as an expression of the urge to transcend the awareness of personal isolation and separateness. This is the same impulse that drives religious worship, military conquest, chemical dependency, and artistic creation, he believes. In romantic love, the sense of being isolated and alone is overcome through a kind of symbiotic union with another, which initially takes the form of *dependency* on the Other, but can mature into a mutual relation of autonomous giving. The problem with Fromm takes the form of sexism and heterosexism, as in his claim that "In the act of loving, of giving myself, in the act of penetrating the other person, I find myself, I discover myself, I discover us both, I discover man" (1956, 31). Although failing entirely to think in a gender-neutral way, he suggests the process of self-transcendence that should be considered (as I have previously argued). After this passage, he writes (still phallocentrically):

> The longing to know ourselves and to know our fellow man has been expressed in the Delphic motto "Know thyself." It is the mainspring of all psychology. But inasmuch as the desire is to know all of man, his innermost secret, the desire can never be fulfilled in knowledge of the ordinary kind, in knowledge only by thought. Even if we knew a thousand times more of ourselves, we would never reach bottom. We would still remain an enigma to ourselves, as our fellow man would remain an enigma to us. The only way of full knowledge lies in the *act* of love: this act transcends thought, it transcends words. It is the daring plunge into the experience of union. (1956, 31)

By "act of love," Fromm is referring to all actions that arise out of an attitude of love, not simply physical union. Looking past the masculinist assumptions and sexist language (characteristic of this period of psychological writing), we again find the idea that self-awareness develops as the self extends its awareness and commitment beyond the boundaries of the self—in other words, through attachments formed via extensibility.

Adolescence is the phase in an extensible individual's life cycle when it is expected and considered normal that one may fall in love with someone outside of the household. The peculiar character of adolescence as a life phase in the United States and other Western societies, then, is the willingness to tolerate an uncertain engagement with the world, an attachment to the unknown, by part of the household. American parents usually set curfews for their sons and daughters, but do not generally send chaperones or require that meetings occur only under parental surveillance. Nor does society at large pass judgment on young women seen alone with young men and consider them to be "fallen" if they do not subsequently marry those same men (though subcultures, in particular immigrants within the United States, may pass such judgments). Even expectations about the "right" and "wrong" sorts of match are dispersing, with miscegenation laws a thing of the past and black-white marriages in the United States increasing 154 percent between 1980 and 2000 (West 2004).

In the short historical period corresponding not coincidentally to the diffusion of the automobile, the fumbling intimacy, excitement, and tragedy of first love has become a central motif of American culture. For half a century, films, songs, and photographs have been steadily created to celebrate adolescent romantic love. Romantic love has thus become a "myth" of high-modern culture in the Barthesian sense—that is, a symbol that carries the freight of a host of other symbols and resides parasitically on many texts.[4] Love is thus elevated to the status of an ideal. Giddens refers to this ideal as the "pure relationship" (1991, 88–98)—a relationship that exists for the emotional satisfaction of the persons directly involved.

We find in this ideal the shattering of the cycle of reciprocity that characterized premodern societies. In such societies, marriage was often undertaken less to satisfy the persons involved emotionally than to fulfill an obligation to their family, to enact a sacred ritual, or simply to adopt an accepted strategy of economic survival. Marriages could not

4. Of course, because romantic love survives in a minority of American marriages (only half of marriages survive, and certainly not all of those maintain the spark of romantic love), we must recognize that this peculiar variety of love has a mythical quality to it in the more literal sense (i.e., it is for many a fiction).

be annulled except in extraordinary circumstances, involving great dis-honor, at least to the woman, in large part because she was a kind of cur-rency in interfamilial relations of circular reciprocity. This pattern still constitutes the rule in many parts of the world. The modern marriage, based on the ideal of a pure relationship, accepts the legitimacy of indi-vidual desire as "a codifying force organising the character of the sexual relationship" (Giddens 1991, 91).

In stable distanciated societies such as those of western Europe, where extensibility has been prized for several generations, romantic love is also prized. Conversely, marriages based primarily on duty and responsibility to one's family are prevalent in place-based societies with low levels of extensibility or in societies just moving into distanciation. This situation is evident when one examines the main points of disrup-tion among immigrants from Asia to the United States. Second-generation immigrants from India to the United States cite conflict over dating and marriage preference as the *main* source of conflict with their parents (Dasgupta 1989; Kar et al. 1995–96). Such conflicts reveal as well as anything the disruptiveness of the encounter between place-based and extensible identity constructs.

Treating romantic love—an intransigently *personal* view of social space—as a basis of major life decisions signifies a peculiarly modern relationship to space and time. Where self-determination presumes ex-tensibility (i.e., in a society such as the United States), the community in which one "moves" as a self-determined agent is never the same as that of one's parents or of any other single individual. Maturity itself is closely tied to mobility (primarily by acquisition of a driver's license) and to distanciated communication (the right to use the phone or Internet at will). Young lovers may go cruising, drive to hangouts, or meet in virtual contexts such as newspaper ads and the Internet. These moments indi-cate not only a geography of places for romantic encounters (both physical and virtual), but a geography of the persons and places one joins secondhand as a potential "significant other," thereby increasing one's geography of connections—one's extensibility. Romantic love be-comes therefore a uniquely modern rite of passage into adulthood: one reaches out into the world to find a unique partner, an unfamiliar Other who serves as a point of entry into a world that is often beyond one's previous horizons of knowledge and awareness.[5]

In contrast, an arranged marriage is a sacrament binding two families together—and it makes sense only if the members of extended families live close enough together to maintain regular contact. Villages rather than cities are the original spaces of arranged marriage. Where the tradition persists, as among second-generation Indian immigrants to the United States, it loses its meaning and social function because the two families brought together by marriage are nonetheless separated by geography—if not from each other, then at least from the new couple. This place-based institution, like many others, becomes fragile and subject to erosion by the place-transcending model associated with romantic love.

This is not to say that one involves control and the other does not: two forms of control are implied. To choose someone "out there" freely and to transform the sense of contingency associated with "the Other" in general into a sense of necessity, "the One," is to bring some sense of stability to an uncomfortably fluid social space. To "fall in love" therefore implies that the individual domesticates the world. Conversely, to marry the person chosen by one's family is to have oneself domesticated, to deny desires that are wild or unruly, and to submit to parents' authority and beyond that the authority of a formula based on factors such as caste, profession, astrological sign, family standing, religion, and place of origin.

As an aspect of distanciated identity, romantic love is not confined to adolescence or to a single rite of passage. In highly distanciated societies, romantic love recurs for many individuals sporadically from adolescence to old age, whether they remain single, divorce and remarry, or pursue extramarital affairs. Building extensibility is almost certainly not

5. The popular film *My Big Fat Greek Wedding* (2002) plays with these themes by showing the generation gap between first- and second-generation immigrants within a society where reaching maturity implies personal autonomy and the supreme expression of autonomy in romantic love. Other films such as *Fiddler on the Roof* (1971) and *Eat Drink Man Woman* (1994, by Taiwanese director Ang Lee) have addressed the same issues.

the lover's conscious goal; at the subconscious level, however, failure or success in romance for an extensible individual can provide a bellwether of success for the lover as an extensible individual more generally (Gergen 1991). Love represents the dream of self-transcendence (tragic though that dream may often be). "Heartbreak" is emotionally disruptive not simply because it throws one *into* the wide world and out of the "home" of a close attachment, but also because it marks a slippage *away from* the wide world (as a dream of pleasurable self-transcendence embodied in a particular Other) and back to the drab confines of one's familiar world.

Anthony Giddens finds that the process of the "active intervention and transformation" in one's self-development that is characteristic of high modernity is particularly apparent in situations of relationship dissolution, such as divorce (1991, 12). But even as individuals are confronted by the risk of losing social connections and networks routed through their partners' social links, they reach out to society at large for advice on how to navigate their difficulties. Their reconstruction of self often involves a host of support structures ranging from self-help books to counseling services, each with audiences and clients at various geographical scales.

In success as in failure, then, romantic love is a spatial process. A lover is not (merely) a master or a possession (which would demonstrate a relation of dominance and submission rather than of love), but also a doorway onto new aspects of the world. A lover demands expansion of one's character—reconfiguration of self not simply to tolerate, but to appreciate differences between one's own lifepath and another individual's lifepath. This is true even if the steps and phases of courtship are bound by unspoken rituals and if heterosexual relations and gender roles are promoted by culture in countless ways that limit and channel individual freedom in the expression of romantic love. Absolute freedom need not exist in order for love (or any sentiment) to entail an avenue for agency.

Of course, the body is central to the expression of many sorts of love, especially romantic love, and that is where one encounters the main battlefields with regard to "right" and "wrong" sorts of love. Although love is thought to be a matter of individual, private action, an older model that places physical contact under community control still manifests itself in laws restricting sexual activities and in attacks on gay men and women. If distanciated society encourages respect for love, that respect is still curtailed by concepts of propriety developed in pre-modern and early-modern societies.

Romantic love's intensity exists in tension with the goal of a pure re-lationship (Can two people use each other for sexual ends but still treat each other with respect?) and therefore likewise exists in tension with the Christian model of brotherly love and the Greek concepts of *philia* and agape. No one lover can embody the diversity of one's social world, so closure is inherent in romantic love. Two lovers may become blind and deaf to the world they inhabit, and concerns such as the oppression of diamond miners or the pollution caused by rose farming are far from their dazzled minds. Asymmetries in power (perhaps rooted in the phys-iological difference between male and female bodies, but embellished through the "amplification" of symbolism) also plague romantic love and often prevent it from reaching the ideal of the pure relationship. So friendship and brotherly love even more closely approximate a Haber-masian "pure relationship." Jonathan Crossan associates Christian brotherly love with a "universal peasant dream of a just and equal world," which means a radical egalitarianism, akin to communism (1994, 74). Brotherly love attempts to transcend the distance separating one person's interests from another person's interests and rejects dis-tinctions based on race, class, or any other objective criterion. Brotherly love is revealed mainly through actions revealing a sense of moral re-sponsibility to strangers—in a word, through *altruism*.

Altruism

An impersonal but specific expression of love, consonant with *philia,* occurs through acts of altruism. Admittedly, altruistic behavior is rare, but it is not so rare that its occurrence in distanciated society causes us great surprise. The countless small acts of kindness in daily life indi-cate that if the cost in time and effort to help another is low and his or her need appears justified, an altruistic response is not uncommon (Tuan 1986b). We can ask directions on a city street and expect a stranger to answer with good intent, if not always accurate information. We can count on someone opening a door if our arms are full or calling the police if we are lying unconscious on the ground. Even extravagant acts, such as rushing into a burning building to save the occupants, are common enough to be featured regularly in the tabloid news. In the framework introduced earlier, we can interpret these actions as a manifestation of extensibility: when the self is *accustomed* to acting and sensing outside of the confines of the body, then another's need prompts action without much reflection about remuneration or reward. Such acts of altruism are not and should not be justified philosophically. Their cause lies in the habit of living outside oneself.

For decades, social critics have decried the withering of social life, including not only formal institutions but informal signs of civic pride. The jeremiads of the 1970s, such as Richard Sennett's *The Fall of Public Man* (1974) and Christopher Lasch's *The Culture of Narcissism* (1978) sounded a death knell for civil society and all its fruits, from small acts of generosity to large acts of altruism. Yet more than twenty years later, people still help strangers and work together to respond to crises.[6]

Altruistic behavior persists in late-modern societies because it is a confirmation of an extensible self-identity. One takes a risk to save a stranger because willingly allowing another to suffer or perish conflicts with the model of self deployed constantly in ordinary life—a model that spills over into relationships with certain strangers. In short, the habit of living outside oneself has a moral dimension, though the limits of this morality are generally defined by the limits one places on "us" and "them." Exclusive definitions of "us"—whether sexist, racist, or classist—often direct and limit altruism, but they do not make it irrelevant as an aspect of the extensible personality.

Altruism depends on living among strangers, like a cosmopolitan or the resident of any large city. This is because a good deed to a non-

6. For example, during a massive electrical failure that affected the northeastern United States and parts of Canada on August 14, 2003, killings, break-ins, and felonies actually decreased, while ordinary citizens began directing traffic (Cruz 2003).

stranger can always be seen as incurring a tacit or implicit obligation. Where no one *is* a stranger (as in a small town or village), it is questionable whether altruism truly exists because "selfless" behavior can always be expected to incur an obligation, so altruism is an investment in the goodwill of others—in short, a form of insurance. Regular encounters with strangers—opportunities for altruism—are therefore much more common in large societies than in small communities, and, given the trajectory toward globalization, we have reason to envision a world in which opportunities for altruism will continue proliferating. A growing range of Others are certainly subject to the perception of the need for assistance, which precipitates a gesture of support, though knowledge of these needy others may be shrugged off. If extensibility creates the opportunity for altruism, it also presents a challenge that may be met with a rejection of all obligations. Some may retreat from the challenge of extensibility and espouse radical selfishness, such as Ayn Rand, who describes altruism as "moral cannibalism" (1964, 34). Indeed, the greatest threat to global security is not globalization, but the espousal of a moral doctrine that supports economic globalization but discourages a sense of moral obligation to the distant others whose lives are economically intertwined with ours.

To discuss altruism raises sticky theoretical issues. As Tuan (1984) argues, we generally love those persons and things that are in some sense dependent on us. By saving another, we bring him or her (or it) into our sphere of affection. Our goodwill is not untainted by power, but it is still greatly preferable to an artificially bounded sense of self. Altruism helps create a kind of symmetry—a congruity of scales—between one's moral world and the space affected by one's actions. This symmetry, in turn, compensates for the asymmetry of power and access. Therefore, altruism is essential to building a humane distanciated society.

Creativity

In an economy dominated by services and information processing, jobs increasingly demand creativity. David Brooks lists forms of institutionalized creativity building in current work environments: "learning

Woodstock" sessions at Xerox, a Joy Committee at Ben and Jerry's, and "humor rooms" at Kodak. Elsewhere we can find "corporate jesters," a "thought theater," and a traveling creativity-building seminar called "orbit the giant hairball" (2000, 130). The promotion of creativity is not confined to the information-age workplace, however. Anthony Giddens argues that a fundamental trait of "post-traditional" identity is the "reflexive project of the self"—the "process whereby self-identity is constituted by the reflexive ordering of self-narratives" (1991, 244). This slightly cumbersome terminology can be clarified by folding creativity into the constitution of the extensible self.

What makes traditional societies traditional is not a lack of change. Even the most traditional societies invent new traditions from time to time. The rate of innovation varies from society to society, but what makes a society nontraditional is more, even, than the rate of innovation: it is *the emphasis placed on personal creativity in the shaping of each individual life.* As David Harvey admits, "The incredible power of capitalism as a social system lies in its capacity to mobilize the multiple imaginaries of entrepreneurs, financiers, developers, artists, architects, and even state planners and bureaucrats (and a whole host of others including, of course, the ordinary laborer) to engage in material activities that keep the system reproducing itself, albeit on an expanding scale" (2000, 204). Harvey means this statement as a warning about the power of capital, but if creativity is in itself a source of enjoyment and fulfillment, then many people in capitalist society have the opportunity to enjoy that luxury.

A person born into a traditional, low-distanciation society has few opportunities to approach life as a creative project and is encouraged not to innovate in matters of identity and life goals. This is especially true of members of subordinated groups: women, ethnic minorities, slaves. But even free men may find themselves with only a few options of action when it comes to things that really matter. Tradition often means one is obliged to accept a predetermined biography, a role dictated by such characteristics as family background, gender, appearance, and place of birth. The biography of the "important" person in premodern societies is often as caging as the biography of the "unimportant." The shift away from tradition affects the goals and products of action as well as the character of action as an expression of individuality. Whether individualistic biographies are encouraged or discouraged for anyone reflects a society's attitude toward creativity in general.

Therefore, insofar as individuals are allowed and encouraged to act creatively to construct a unique biography, a society may be seen as nontraditional. No society is entirely free of the influence of roles given in advance by family background, gender, appearance, and place of birth, but these roles are less confining and more often subject to revision in nontraditional societies than in traditional ones. At the historical point of transition, multiple traditions vie for attention, and there is no unanimously endorsed biographical model of the self. People are confronted daily with a diverse array of lifestyles that serve as alternative models for their own future development. Women's roles loosen up more slowly than men's roles and may temporarily suffer greater constraint; a society that welcomes women in all social roles is one that places creativity unequivocally above brute force and aggression. This is not to say that women (or men) thereby escape the operation of power, but at least they are better able to define personal goals and to achieve them.

Giddens puts it in simple terms: in high modernity, we are "not what we are, but what we make of ourselves" (1991, 75). We can "make ourselves" in ways that are more or less creative, though all ways are directed to a degree by the capitalist system, and it is clear that people differ in their inherent level of creativity and thus in the distinctiveness of the identities they create. Many people in distanciated societies simply adopt a convenient model from the media or from family. Richard Sennett (1974) suggests that media models take the form of popular personalities with a kind of stature that ordinary people cannot possibly acquire, and by emulating these personalities (actors, actresses, talk show hosts, models, performing artists), people surrender their individuality and become conformists. Ironically, the "cult of personality" arising from the constant reinvention of self can lead to conformity. But even the conventional choice is a choice as long as there exists the possibility of being and doing something more daring.

The transition toward creative construction of self has been in progress for five centuries in Europe. Shakespeare's wry wit reflects the

European Renaissance, a time when social roles of all kinds were already beginning to become more fluid. His character Jaques in *As You Like It* offers an image of the life cycle that is a parody of the life dictated by convention.

> All the world's a stage,
> And all the men and women merely players:
> They have their exits and their entrances,
> And one man in his time plays many parts,
> His acts being seven ages. At first the infant,
> Mewling and puking in the nurse's arms.
> Then the whining school-boy, with his satchel
> And shining morning face, creeping like snail
> Unwillingly to school. And then the lover,
> Sighing like furnace, with a woeful ballad
> Made to his mistress' eyebrow. Then a soldier,
> Full of strange oaths, and bearded like the pard,
> Jealous in honour, sudden, and quick in quarrel,
> Seeking the bubble reputation
> Even in the cannon's mouth. And then the justice,
> In fair round belly with good capon lin'd,
> With eyes severe, and beard of formal cut,
> Full of wise saws and modern instances;
> And so he plays his part. The sixth age shifts
> Into the lean and slipper'd pantaloon,
> With spectacles on nose and pouch on side;
> His youthful hose, well sav'd, a world too wide
> For his shrunk shank; and his big manly voice,
> Turning again toward childish treble, pipes
> And whistles in his sound. Last scene of all,
> That ends this strange eventful history,
> Is second childishness, and mere oblivion,
> Sans teeth, sans eyes, sans taste, sans everything.
> (1975, II.vii.139–66)

This mockery of social roles, echoed by various forms of masquerade in Shakespearean theater, signals a period of social transition. The list of life phases was meant in jest, and the jest was possible because with all the increase of specialization and autonomy in early-modern English society, any roles familiar enough to be described lent themselves naturally to ridicule. The biographical sketch is not a model, but a caricature that only a few people might approximate, yet it delights because of the obvious irony behind the appearance of universality. Contrast this description with the Hindu conception of the life phases, called *Ashrama:* "*Brahmacharya,* the stage of the celibate learner; *Grahasta,* the stage of the householder; *Vanaprastha,* the stage of gradual disengagement from worldly duties and loosening of social bonds; and *Sanyasin,* the stage of complete disengagement leading to renunciation for achievement of spiritual freedom" (Mishra 2000, 178).

This model is not written to provoke wry laughter. It is understood as a guide to life by a community grounded in tradition. *Ashrama* is a kind of template for one's life, not a commentary on the way people end up willy-nilly living their lives. When models are treated with levity more than seriousness, as in Western cultural products from Chaucer to Shakespeare to *The Simpsons* to *Married with Children,* this ironic treatment of life roles indicates that some measure of creativity is demanded of a person.

The objects of creativity in premodern societies were mainly material artifacts—weavings, baskets, pots, clothing, ornaments, religious icons, dwellings, and so on—and one did not need to be a creative basket maker in any case. The objects of creative agency in "post-traditional" society are selves, and one *must* actively create a self in order to exist. A link can be drawn, then, with the theme of love: it is precisely when all the rules of improvisational self-construction are changed through a momentous act such as declaring or acknowledging one's love for another that the communicator suffers a crisis of identity, feels a sense of uncertainty as to how best to continue to create him or herself, can no longer find appropriate words, and begins to sigh "like furnace." Communication in words fails precisely because extensibility has bounded ahead of identity.

Of course, not only the self, but the environment and all living and nonliving elements of it are the normal field of creative action. Creative skill of this sort has a distinct economic advantage for those individuals who are unusually creative. As quaternary (service) activities come to dominate the other economic sectors in terms of employment volume, creativity becomes ever more important to economic expansion, and those persons who are unusually creative can cash in grandly on the market for creative people. Conversely, in more local and traditional societies, people who demonstrate too much creativity (particularly in the construction of identity and lifestyle) were and are subject to scorn, ridicule, and ostracism. In distanciated society, creativity is rewarded with explicit protocols that bring recognition to the creative individual: contests, competitions, certifications, copyrights, patents, privacy laws, and the like. Although not all of these modes of recognition are designed to encourage creativity, they all assist it in one way or another.

From a communicational standpoint, one of the most interesting of these protocols is the emergence of authorship as a basis of authority. As Marshall McLuhan (1962) observes, artistic, musical, and literary creations often were anonymous prior to the invention of the printing press; the rise of mass literacy paralleled the rise of the "author." Earlier in European history, the stable economic system, the overwhelmingly local basis of community, and the centrality of the church discouraged attempts to derive personal glory from one's artistic or intellectual creations. Of course, this meant that artistic and intellectual creativity was not widespread. With the mass production of words and images, beginning shortly after the invention of the printing press and accelerating sharply from the eighteenth century on, every person could be seen as a potential author, a creator with a unique public identity linked to his or her unique creations (Eisenstein 1979, 152–59).

Ironically, even as creativity became a marketable commodity, with the most talented musicians, artists, computer programmers, and screen actors able to earn tens of millions of dollars, the market began a slow attack on the creativity involved in the nonwork spheres of society. Mass production and mass marketing meant that fewer and fewer people needed to develop in themselves the skills for creatively passing the time. An adult who regularly plays an instrument or paints pictures is an anachronism. Home-made clothing and other material goods are increasingly shabby by comparison with the sophisticated products of a high-tech, highly automated, computerized industrial complex. The difficulty in meeting the high standards of machine-tooled products means creativity is only rarely channeled into the making of everyday objects, an avocation that once dominated human creative efforts.

Miguel de Oliver and Michael Yoder argue, for example, that residents of their San Antonio study site, "like most other Americans, will continue to fall, one by one, into the abyss of self-centered individualism expressed in material symbols. This individuality represents a more lucrative market for producers and only reflects a diversity of commodities rather than a diversity of culture" (2000, 105). We must not overstate the threat to creativity by overlooking the production of a biography as a creative process or the great demand for creativity in the quaternary workplace, but clearly a tension exists between the economic necessity to be creative and the commodification of ever-widening spheres of experience, which saps the creative spirit in non-work activities.

The Challenge to Extensible Identity

Can we become trusting, loving, creative individuals? That is the challenge put to us by our distanciated social context. As Lasch, Sennett, Fromm, and Gergen have shown, it is a formidable challenge. Just as economic systems (e.g., capitalism) and political systems (e.g., democracy) must be critiqued and reconceived, so too must our ideological frameworks. Does our worldview promote creative, trusting, loving engagement with the world? Or does it conversely encourage people to withdraw into a shell of bitter cynicism, wherein love, trust, and creativity appear to serve no purpose and therefore are ignored? It is far too easy to adopt the pessimistic route and thereby close off the path to a more equitable society in which the spaces of agency and moral commitment are symmetrical. Marxian theory and "critical theory" in particular promote this sort of pessimism.

If we see extensibility as only a means to an end, in particular as only a means to making money, then we are likely to consider communication quite cynically. If we understand that extensibility is something people desire for its own sake and for the sake of a community, as a form of power that enables people to work for a society where love, trust, altruism, and creativity are valued, we will see the situation more clearly. In this light, the training in extensibility that people receive from the time they are children may solidify their class advantages, but it may also provide them with tools to overcome their biases and to construct a just society.

If we turn from education to mass-communication media (two sources of enculturation often construed as apparatuses of ideological hegemony), the essential implication of what I have presented is that insofar as people use media as means of extending themselves through space, this form of extensibility (like all expressions of human agency) can involve working against exploitation and oppression rather than simply reproducing the dominant forms of economic, political, military, and ideological power. But if this point seems to evoke mediated communication as a kind of cultural battleground, we must recall that the frameworks of trust, love, altruism, and creativity work against exploitation and oppression not by attacking the "exploiters" and "oppressors," but by forming a sense of interpersonal or collective attachment—an awareness of merged agency, overlapping objectives, and the desirability of transcending the narrow confines of self-interest or parochially defined group interests. Although we may be obligated by historical awareness to approach new communication technologies with suspicion, we should be inclined for pragmatic reasons to welcome their potential to expand the bounds of the self.

3

Communication Content

The Foundation of the Self

> [N]o matter what I say, I must employ a language which itself re-
> flects both the world and my view of the world.
>
> —Gunnar Olsson, *Birds in Egg*

MEANING ARISES OUT OF MORE than just the content of a communication; if I say "hello" in a bedroom, office, or lecture hall, in a
note, on the telephone, in an e-mail, the effect of the word is different
each time. Communication *content* is the "hello" itself, whereas *context* is
the medium used to convey the content, considered broadly so as to include not only rooms and other physical structures, but also telephone
systems, computer networks, broadcasting studios, and the social formations that support the creation and maintenance of this communication infrastructure.

Whether people communicate in place or through media, meaning
is a product of a complex interaction between (1) communicators
(which includes senders and receivers), (2) messages and meanings, and
(3) contexts, including physical containers and technological infrastructure. The messages and meanings can be theoretically isolated under the
term *content,* though in reality content always exists in some kind of context, and the context places certain constraints on what can be communicated, so the two are inextricable. For example, both a dance and a
sermon can convey a sense of tragedy, but they do so in very different
ways. Likewise, a book and a movie can tell the same story, but they do

so in different ways. In the latter case, we often think of the film medium as requiring things to be "left out," but if the sequence were reversed and the book came after the movie, we would again speak of things needing to be "left out" (music and sound effects, for example). A certain amount of content is shared between the book and the film, but differences between the book and the film as contexts force differences in content.

Although content and context cannot be separated in fact, this chapter isolates content and chapter 4 isolates context to explore ways of better understanding the *grounded* nature of each of these elements of communication—their roots to place—and their ties to the construction of the extensible individual. Specifically, this chapter introduces content, then applies a theoretical framework based on the sign, the symbol, and the signal. I use these three *relata* to understand the act of walking in the city. Chapter 4 turns to context and applies a framework based on the interplay of material and social structures to understand a "walk" through the Internet. In both chapters, I argue that communication is grounded in a geographical reality that shapes meaning. This material reality forces the continuous revision and adaptation of communications. I also argue that the self is constituted as a specific kind of individual *by communication:* as content is framed in particular contexts, space is created for particular kinds of (extensible) senders and receivers.

Signal, Sign, and Symbol

Communication content consists at the most elementary level of three basic kinds of meaningful units—*signals, signs,* and *symbols*—though they are combined in countless ways.[1] Signs are built out of other signs, symbols are combined to form more overarching symbols, and signals are both the most basic of communications (shared with

---

1. This tripartite scheme is derived from that of Roland Barthes (1967, 35–38), which in turn reflects the distinctions between *relata* identified by Carl Jung, G. W. F. Hegel, and Henri Wallon.

even rather simple animals) *and* the most sophisticated (forming the ultimate "purpose" for vast complexes of human communications such as religions, markets, and philosophies). Like complex molecules built of a finite number of atoms, the meanings circulating in and out of particular places in any given life are combinations of signs, signals, and symbols. Similarly, the concept of self involves the three types of *relata* assembled in a unique and thickly layered way.

If the place in question happens to be a church, then the units of meaning include candles, pews, stained-glass windows, stone sculptures, Bibles, hymnals, steeple(s), bells, a minister, sermons, and rites or sacraments, among other things. These units are brought together in a process of symbolic accretion to form a master text that is the place. Some of the church's constituent elements are arbitrary, such as the word *God* in an engraved text, which, were it rendered as *Dieu* in a French church would mean exactly the same thing. This ability to substitute words indicates that any given word is arbitrary. Other elements are nonarbitrary. They are *motivated,* meaning they can vary in shape only within certain limits because they represent experiential aspects of something, such as the image of the cross. As in architecture, the guiding principle of this type of *relata* is "form follows function." The motivated atoms of meaning (e.g., the cross) are called *symbols,* and the *unmotivated* elements (e.g., the word *church*) are *signs.*

Some of the elements making up the church tell people to *do something* in the here and now (for example, the sound of bells can mean it is time for the service to begin). The word *church* and the symbol of the cross do not work in this way. Calls to act in some way or perform some behavior are called *signals.* Other elements of meaning say something *about* something. A signal, such as the church bells, is not suited to discussing the past or future. Its subject is the present.

Things get complicated, however, when signs, symbols, and signals are assembled into larger units like molecules. Any story with a moral is meant as a signal, although its elements are signs (words) and symbols (tropes and plots). A church, too, is most generally and abstractly a signal because its presence inspires some people to act in a certain way (praying, singing, experiencing faith, guilt, etc.). Any place—a mosque, a

Hindu temple, a Buddhist temple, a city, a house, a university, a shopping mall—is an agglomeration of various forms of *relata* to form a complex heterogeneous entity that, as a signal, urges people to act in some way or perform some behavior. This (usually) silent persuasion can be ignored or resisted, but doing so has consequences for anyone who dares not to act or dares to act "out of place" (Cresswell 1996, 2001) or not to act at all.

All of the elements of meaning—signs, symbols, and signals—have in common the function of creating a *relation* between two things (objects, phenomena, actions, sensations, ideas, forces, substances, structures, systems, processes, etc.). These bits of communication content differ not only in terms of the motivated-unmotivated distinction and whether they tell the listener to *do* something or simply tell *about* something. They also vary in terms of their relation to space-time: signals are tied to the here and now; signs allow structured delineation of here and now, there and then; symbols form connections between here and now, there and then.

Despite this variety, the meaning of communication content is always grounded in place. Not only must we interpret *relata* of all three kinds by using a place, but our active appropriation of space becomes itself a communication with aspects of all three *relata*. In other words, and as shown in detail at the end of this chapter, a walk in the city is a form of utterance (Certeau 1984). That utterance "speaks" to others and to oneself; it says something about being you or being me and something about *being* in general. This is true of all sorts of movement. Landscape design, architecture, streets, vehicles—all come together through the process of human occupation to say something about what it means to be a particular person in this place and time.

Signals

A signal is a call to action that functions in a concrete and immediate way. A traffic signal "says" go, caution, or stop and implicitly tells a viewer when and where to do so. A green light means "you go, right here, right now." A red light means "you stop, right here, right now."

Similarly, a finger placed across one's lips means "you be quiet, right here, right now." In other words, signals are fixed in time and address: they always address the second person (you, the listener, the viewer) and do so in the present tense. All that varies is the indicated action. The fact that their spatiotemporal and social dimensions are inflexible means they can motivate actions in a direct way, and in fact they have a unique kind of power because of this.

Pavlov's well-known experiment conditioned a dog to salivate in response to a bell signal (Pavlov and Anrep 1927). This experiment calls to mind any number of sounds that can cause immediate, conditioned responses in people and animals: the sight of food laid out on a table, the smell of roasted meat, the soft feel of fruit in the palm—any of them can be taken as a signal to eat (though we must be clear that the signal resides in the perceiver's mind and not in the thing itself). Signals can be visual, olfactory, auditory, or even tactile, but in each case the signal is, in essence, a link or association between a sensation and some rather specific action. If the sensation-action connection is instinctive, we have what is called an unconditional response, such as a dog's instinct to salivate at the smell of meat. This response can easily be "rewired," associated with a learned connection to a conditional stimulus (e.g., a bell) to produce a *conditional response* (e.g., salivating when a bell rings).

Signals may be *motivated*, linked in some nonarbitrary way with their meaning, or *unmotivated*. Pavlov's bell was an unmotivated signal because food normally has no connection to the sound of a bell. Similarly, the word *fire* does not sound like the crackle of burning logs. In contrast, the signal of bared canine fangs is motivated: it shows the potential pain of being bitten (a signal easily read by both humans and animals). It is tempting to apply the conditional/unconditional distinction without ambiguity, but many signals fall somewhere in between. The traffic signal is weakly motivated: red carries associations with fire and blood, and hence danger, just as green carries associations with growth and youth, and hence motion. They are rather abstract forms of motivation; in nature, red berries may be poisonous, or they may be ripe and good, so the color has unconditional links that are contradictory, and the motivated character of the red light is not the same as an unconditional response.

Not surprisingly considering Pavlov's experiment, animals' communications consist largely if not entirely of signals: bared fangs, raised fur, squawks, growls, screeches, howls, hoots, bird mating dances, even honeybee navigational dances. Like the signals used by people, these animal signals have no need of syntax, and they elicit an immediate response (though that response may vary depending on the receiving animal's size, sex, age, and other conditions). For animals, such responses fall within a narrow range, including fight, flight, mating, imitation, and, among the "higher" animals, play. Honeybees communicate about nectar sources with ornate dances, but this does not constitute a "discussion" about flowers; they cannot argue about whether clover or buckwheat nectar makes the best honey. People move seamlessly across the divide between signaling and signing, doing something like a honeybee dance (signal) to indicate the unexpected arrival of donuts in the office or a celebrity on the sidewalk, but then breaking into conversation (signs) as soon as the gestural signal to gather has produced a crowd.

We must not exaggerate our separation from animals. Chimpanzees can be taught to use signs and signals that people use (American Sign Language), although their communication almost always deals with the here and now (Fouts 1997). This discovery followed the recognition that chimps are much better equipped to use gesture than verbalization as a communication modality. No doubt, other animals will be added to the list of sign-using animals once people recognize their favored communication modalities, but even chimps, presumably near the top of the animal communication hierarchy, employ signs mainly as if they were signals: commanding, remarking, and begging much more than reflecting or discussing. Furthermore, when people use signals, they often do it in a way that is closely aligned with animal behavior. Sighs, yawns, groans, sobs, snorts, grunts, giggles, and guffaws all work on this level. A baby knows instinctively that crying will signal the mother to satisfy its needs (even before it has any sense of what those needs are), and the instinct remains throughout life. At all ages, people signal each other involuntarily and without being fully aware of the messages they are sending. These signals often erupt through daily interaction as an unwelcome intrusion on the way things are "supposed" to be. A blush, for example, signals embarrassment and involuntarily encourages another to be bolder or less bold, depending on the culture, the sender, the receiver, and the situation. A sharpening or softening of the voice is a signal laminated onto the signs of speech and shows otherwise concealed emotions when tensions mount. It is a remnant of prehistoric responses to tension: a hint of a hoot. So culture defines and directs the meaning of signals, but signals are not fully domesticated by culture.

Geographers became aware starting in the late 1980s that the human awareness of a place depends on the ability to name things and imbue them with symbolic significance (Barnes and Duncan 1992; Cosgrove 1984; Cosgrove and Daniels 1988; Duncan 1990; Duncan and Ley 1993). This literature on place iconography and sense of place is far more attuned to signs and symbols than to signals, and the bias remains to the present. What it overlooks is that there is an inescapably *animal* element in the way people occupy places, a communication behavior grounded in signals. This is why we form such strong bonds with dogs and cats: the cat lolling on the warm doorstep and the dog barking from behind the fence are our alter egos, beings that say to themselves and to the world what we are usually too civilized to say, but what we resonate with at a level beyond words. We do loll on the beach and "bark" at the door-to-door salesperson. Our furry friends therefore *share place* with us in ways that the goldfish in its bowl cannot.

Distinguishing between signals and other relata is not always easy. Drawing on George Herbert Mead, Habermas gives examples of several signals that take the form of verbal exclamations (and hence would seem to be signs): "Dinner!" "Fire!" and "Attack!" These words act as calls to action and are complete utterances in and of themselves (1987, 6). "Fire!" is not a detached observation about the oxidation process, but a call to "you, here, now." Such signals have much more in common with our animal ancestors' warning screeches and hoots than with ordinary human speech. Interestingly, we associate self with sign communication, not with signals, and to be reduced to signals (as when cursing or gasping in pleasure) is understood as temporarily losing track not just of one's signs but of *oneself*.

Considered more objectively, signals are not worth less for being shared with animals, for frequently marking the far side of conscious control, for having fixed qualities relative to time-space, or for "speaking" about a narrow range of alternatives (eat or go hungry, live or die, mate or remain solitary, etc.). The fact that most human communications taken in their entirety add up to signals indicates that the signal remains essential to human life.

The importance of the signal is also revealed by its variety in modern culture. "Stop," for example, can be signaled by a stop sign, a stop light, a red automobile tail light, an outstretched hand, an outstretched fist, a shake of the head, a spoken word, a shouted word, a fence, a railing, an alarm buzzer, a blush, a tear, or a particular kind of silence. Each is at one time or another terribly important to someone. *Which* signal we use in a place is itself a signal of a more elusive kind. A stop sign is in fact a signal, and as such it does not simply signal "stop"; it signals that a street carries a certain amount of traffic, not as much as a street with stop lights and certainly not as much as a street (freeway) that has no stops at all. We find, therefore, that physical contexts, places, are largely defined by a particular combination or concatenation of signals.

Not all signals are "sent" by people. Some are inherent in a situation. These signals might be called "environmental signals." If my car tire goes flat, the first indication I have of this may be a hiss, or the steering wheel may pull to one side. The former is an auditory signal, and the latter is a tactile or haptic signal. Either one says, "You, here, now, handle this car differently." Without a doubt, I receive a signal when one of these things occurs, and the signal causes me to act. But who sent me the signal? Likewise, if a funnel cloud forms overhead, I take it as a signal to move into the basement or under a bed. How can machines and even nature itself signal to us if they are not conscious agents? The answer is that signals are not "things" transmitted from one actor to another. *They are temporary states or attributes of a connection between things.* Insofar as I am connected to my car by driving it and to nature by living in it, my own agency is extended out into the car and nature. The signal I receive from the car is in fact a change in the status of a part of my complex environmental linkages, which is to say a part of myself.

Signals occupy the links that make up the network of society, where they come and go like fish in a river. Some flicker by for a few moments. Others are permanent changes in a social link: a love letter, a "Dear John" letter, a contract, a court order. Environmental signals, such as flat tires, funnel clouds, or symptoms of illness, are also changes. They affect human-environment links and social links.

Signals simultaneously say something about the human and the physical environment the human inhabits. Signals are one of the main ways that environments "make sense" to people. The arrangement of furniture in a room signals the room's use, whether it is an office or bedroom, a place for dancing or a place for eating. A podium at one end of the room signals to the occupants that one occupant will be addressing the rest of the room's occupants and that when this occurs, the occupants should face the podium. Beyond this superficial interpretation, the podium signals that one speaker of the entire gathered assembly has a special authority to speak. *Why* this is so and *how* it can be so take us to the realm of symbols and signs.

Our first encounter with the inseparability of context and content has been in the realm of signals. We have seen that signals are fundamental aspects of environmental relations that humans share with animals. Places may not shape their users, but users shape themselves subconsciously according to the signal logic of places. As atoms of meaning, signals are not, after all, things, but rather states of the connections that people form with things, environments, animals, and other people. Signals occupy the here and now, and in doing so they largely define the character of the here and now. We learn them without always knowing it, and we often respond to them without reflection. The signal-response loop is part of who we are, as deep as animal behavior, and it often strains against the rule-based systems of signs and the abstract associations of symbols.

Signs

Signs are conventional, arbitrary associations between particular meanings *(signifieds)* and particular visual or auditory stimuli *(signifiers)*. In speech, the signifier (also called the "sign vehicle" [Janis 1965, 55]) takes the form of a sound pattern, or rather a small collection of sound patterns recognized as semantically identical. The theory of sign systems, called semiotic theory, posits that signs are not sounds linked to *things;* instead, signs are patterns (patterns of sound, patterns of ink on the page, etc.) linked to concepts (mental models of objects, phenomena, relationships, etc.) (Barthes 1967; Eco 1976; Saussure 1983). The word *creek,* for example, consists of an arbitrary link between a set of phonemes, \krék\ or \krik\, and a set of objects (small, running water bodies). The signifier is a combination of sounds that varies somewhat from speaker to speaker but remains within very specific bounds so as to avoid confusion with other words such as *crook* and *clique.* The same signified may be specified by completely different patterns, such as the spoken word's sound pattern and the written word's shape pattern.

In a famous metaphor, Ferdinand de Saussure explained that signs are like sheets of paper, with figures on two sides that are linked arbitrarily and yet, for the speaker of a language, inseparably (1983, 66–70). Perception absorbs the classifications from language so that both encoding and decoding come as second nature to those with ability in a language. When I look at a house, I do not think, "That thing is called a house"; I simply know it *is* a house, as if the name were attached naturally to the thing. What is actually signified by "a house" is not a thing, a discrete and naturally identifiable object or phenomenon, but rather an extraction of certain experiences from the flow of experience: that set of signals that we agree to call a house rather than a hut or a cabin or a condominium.

In French, the word for jellyfish is *la méduse,* and shellfish are *les fruits de mer* (fruits of the sea) or *coquillage* (shelled things) when the reference is not culinary. Neither term indicates a conceptual link to fish *(les poissons)* that is found in the English words *jellyfish* and *shellfish.* The French call the shark *le requin,* and the eel *l'anguille,* indicating in neither case that the animal is a fish. But English is no more helpful about indicating that sharks and eels are fish. Simply stated, a fish (a chordate with internal gills) is not always called *fish,* and what *is* called *fish (poisson)* is not always a fish *(poisson).* Two fishermen who speak English may debate about whether the better fish are found in a creek or a stream, but a French fisherman would be mystified by this debate because either creek or stream, in French, is *ruisseau*. The invisible lines that divide house from nonhouse, fish from nonfish, stream from creek are not given by nature itself and do not remain unchanged as we cross the boundaries between sign systems (languages). Words are tools that bring order to the world, but that order is never automatic or natural. As we learn a language, that language's order becomes an aspect of self that we share with other speakers of the same language. Our thinking is brought into line, but as Gunnar Olsson points out, "To categorize is to fetter. Not to categorize is to tear the world asunder" (1975, 94). Words seldom stand alone (and never do so *as signs*). Words exist in statements, texts, and discourses that impose order on arbitrary units while indicating the overarching order of time and space in a particular culture and subculture. As Olsson has demonstrated, though terms may be clearly defined in the causal inferences of science, their meaning in practical reasoning is more fluid and context dependent. So the discourses in which words are embedded constantly inflect their meanings and implications even as words impose order on the world. To a certain degree, this observation implies that speaking is a signal even as it involves the use of signs: "To understand a language is to understand that words function both as labels and as signals" (Olsson 1975, 26). If as a guest in a friend's house I say, "I like what you've done with the living room," I am signaling friendship in a general way, and my intent to send this signal will overrule the temptation to add that I think the couch looks like a rhinoceros. Still, in using words, I am forced to create at least the appearance of a different message, which is the nature of the "white lie."

Neither *house* nor *creek* can stand alone as a meaningful statement (*fish* might stand alone, but in that case it would become a signal meaning "you here, now, it is time to fish"). Signs require a linear sequence of preceding and following words plus conversion rules (syntax) linking the one-dimensional sequence of words to a four-dimensional space-time, plus some kind of framework (such as causal inference or practical reasoning) to make sense. The layer of meaning conferred by word sequence can be called "syntagmatic" meaning (Saussure 1983, 126–32).

The internalization of arbitrary links between each signifier and its carefully delimited range of signified phenomena as well as the associated syntagmatic and inferential rules for stringing words together to mean this or that four-dimensional relationship allow human experience to be radically different from the here-now of animals with their signal languages. Chimps may occasionally cross the line, as perhaps do elephants and dolphins, but what makes human communication different from most animal communication (including that of dogs and cats) is the overwhelming reliance on signs and their "space."

Sign systems constitute a peculiarly human relationship to space-time. This relationship is one that ranges widely in time and space through the use of the imagination and the exchange of information with others; in short, it presupposes and supports an extensible self. We might call the four-dimensional space-time of the sign text a "virtual space," but that does not mean it is not real. What makes any space real is that it serves as a context in which people coordinate actions and undertake complex projects. Not only does language do this, but it also permeates our experience of what is more commonly called the "real" world—that is, physical space—giving that world shape and substance.

Words not only *can* indicate relations of time, space, and human actors, but *must* do so if they are to make sense. This sets them apart from signals and symbols. "Run" or "running" alone is not a complete thought because it is too ambiguous. "He was running," "you will be running," "I ran to school," "the car still isn't running," "she will run for president" are all meaningful statements because some relation to time or space or particular actors is syntagmatically created. This virtual space of syntagmatic relations allows the listener to ground himself or herself in an organizing matrix of space and time. Using words to construct space-time relationships is not an option; it is required of sign users. This aspect of sign communication differs markedly from signals, which, as we have seen, are anchored in the here and now. It also differs from symbols, which, as we will see, build links between past, present, and future in a powerful but vague way.

Each language, then, creates its own world, demarcating phenomena and arranging them in precise fashion in space and time (Whorf

1956, 102–24, 199–206). The relative similarity of languages in a family, as among the Indo-European languages, creates a certain parallel in the underlying spatiotemporal perceptual systems of the persons accultur- ated in these languages. For speakers of these languages to learn non- Indo-European languages, such as Hopi, Chinese, or Arabic, necessarily implies a partial relearning of space, time, objects, relationships, and causality. The degree to which a language shapes the "higher" functions of the mind such as problem solving, creativity, and moral and aesthetic sensibility, however, remains an open and intriguing question (Lucy 1992a, 1992b).

Benjamin Whorf points out that the sentence "I went down there just in order to see Jack" contains "only one fixed, concrete reference: namely, 'Jack' " (1956, 259). The rest of the words take on their meaning through their position in the sentence and through the overall pattern created by the placement of the other words: "even 'see' obviously does not mean what one might suppose, namely, to receive a visual image" (1956, 259). Context helps us to understand that to "see" Jack means to have some kind of interaction with Jack (which would not be the case if one went to "see" a movie). This context involves not only the syntag- matic context of the word in the sentence, but also the social context of the articulation in a particular society. Let us say the phrase is spoken in protest by a man whose wife has just accused him of having an affair. Then "I went down there just in order to see Jack" is an assertion that nothing else (like a visit to Ursula) happened on the trip in question. Per- haps, however, the speaker ended up winning the lottery with a ticket purchased at a gas station in a distant town, and the improbability of this event prompting the comment "I went down there just in order to see Jack" in this case would carry a certain irony (signaled among English speakers by a faint curve of the lips and raised eyebrows).

These examples demonstrate that although the language syntag- matically marks out a virtual space and time, it depends heavily on phys- ical space and time—the world occupied by bodies—to make meaning. Yet languages also bear traces of the physical contexts in which they have evolved.

When English-speaking people settled in the area ceded to the

United States by Mexico in 1848, they continued to use Spanish words for certain geological formations: *mesa, arroyo, playa.* A mesa is a flat-topped mountain, an arroyo is a riverbed that is frequently dry, and a playa is a lake bed that is frequently dry. Such physical features were rare to nonexistent in England, so the English language lacked words for these features. Not surprisingly, the language that did have words for them had evolved in a place with an abundance of arid landforms—Spain. (This is of course not to say that nature dictated that a dry riverbed must be called an *arroyo.*) Once English speakers entered a drier, more rugged environment, English was expanded through borrowing so as to better "map" the terrain of experience, though the signifiers adopted for this purpose could have taken an infinite range of forms.

Content reaches far beyond physical environments. My vocabulary and grammar reflect not only my linguistic community (those who speak my language) and subcommunity (those who speak the same dialect), but also my profession, literary exposure, travels, education, interests, and so on, even my family's medical history. Every person's lifepath has provided the opportunity to acquire specialized lexicons and particular world models. The meanings I attribute to words are traces of the places where I have lived. Growing up in Colorado, a state that is both high in altitude and semiarid, I saw creeks small enough to step over. The wider creeks were often shallow enough to cross by stepping from boulder to boulder. When I think of a "creek" now, I still think of shallow clear water between narrow banks, running over stones, and making lots of noise. If I had grown up in the Midwest, it is unlikely that the word *creek* would evoke these images for me. Although recognizing this environmental connection, we must not lose sight of the fact that sign-based communication is a political process because its apparent authority may be resisted or disrupted by identifying new and unexpected terms. This political contestation often boils down to generating third terms that destabilize neat binary dichotomies (Soja 1996). Many binaries employed in sign communication seem to demand our acquiescence to a preestablished way of thinking, seeing, and doing things. Is it (or is it not) a nice neighborhood? Is parking at the store adequate (or inadequate)? Is the trip easy enough (or too difficult)? In these examples, the implicit dichotomies of *nice, adequate,* and *easy* conceal great complexity. *Nice,* when linked to neighborhood, bundles up class prejudice, and *adequate,* when linked to parking, presumes that everyone who shops at a store should be able to come by automobile. As shown in the next chapter, the root of this situation is not the sign or its syntax but an overarching context: unequal power relations structuring a particular social system and its communicating members.

Some general characteristics of sign communication shape relations to physical and social contexts. In historical perspective, signs have careers that rise and fall as the realities they refer to change, as society changes, as understanding of the world changes, and as the sign itself affects society. The French social theorist Jean Baudrillard explains this change in terms of a "procession of simulacra" that leads eventually to the deterioration and loss of the signified. The "hyperreal" replaces the real as the function of the sign passes through four stages: (1) "it is the reflection of a basic reality," (2) "it masks and perverts a basic reality," (3) "it masks the *absence* of a basic reality," and finally (4) "it bears no relation to any reality whatever: it is its own pure simulacrum" (1983, 11, emphasis in original). We can represent each of the four stages with the place-name Riverside. Initially, a prestigious streetcar suburb (near Chicago) is given this name because it really is located alongside a river. As the area develops, the "Riverside" identity spills out through space, lending prestige to a sector of the city that is farther from the river, perverting "Riverside" as a descriptive label. Later, suburbs in other regions are given the same name, even when no rivers of note are visible in these areas, simply because the name carries a certain prestige attached to its earlier connotations (wealth and the elite ability to dwell close to nature). The name has become a mask hiding the absence of a river and, beyond this, the accelerating destruction of natural environments. Finally, people lose all expectation that a place named Riverside, Lakeside, Lakeview, Lakeline, or Shoreview will have any relation to water, and they either lose all interest in rivers or look outside the boundaries of such suburbs for aquatic experiences. People are not surprised that laying out a place called Riverside entails burying a river in a concrete conduit or that building a place called Pine Hills involves the removal of pines and the planting of grass. The suburb loses all ties to nature except what is wrapped up in the sign itself, and the resultant sprawl displaces natural ecosystems, so real nature is displaced by a hyperreal nature.

This is sadly the final moment in the procession of simulacra, and Baudrillard does not sketch a route either beyond this exhaustion of the sign or back toward a more meaningful use of signs. Still, we should note that although (nonlinguistic) reality appears in Baudrillard's theory as something left behind in the procession of simulacra, its ontological status as real and prior to language is not in question, nor is its *potential to be signified* because this tie to reality is the pristine condition in the first stage of his procession and presumably his reason for critiquing the procession of simulacra. Baudrillard carries his scheme to the point of absurdity, however, when he argues that the procession of simulacra has reached a stage in which life now dissolves into television, causality is extinct, and nothing exists but what is designed for infinite replication, like a videotape or a tract house (1983, 55, 146).

Communication can be evacuated of its ties to the real world only by unanimous consent combined with the acquiescence of natural systems such as weather and ecosystem dynamics. As much as people might want to live in a simulacrum, nature itself will not cooperate. Environmental catastrophes often make people rethink their representations of the world. When parts of Kansas, Oklahoma, Texas, and Colorado refused to "behave" as farmland, the arid soil, loosened by the plow, turned into a "dust bowl." The situation prompted various governmental interventions aimed at preventing a recurrence of the phenomenon. The prairies of this region might have looked superficially like other prairies, but they were different. Lines needed to be drawn in terms of policy if people were to continue to live on this land. Familiar terms such as *soil conservation* and largely forgotten words such as *snuster* entered the English lexicon (Bonnifield 1979).

Baudrillard's procession of simulacra suggests the gradual decline of sign-based communication into meaninglessness, but we can just as easily envision the proliferation of signs and sign-based meaning. Saussure's (1983) semiology (also called semiotics) sought to analyze a wide range of cultural domains, "symbolic rites, forms of politeness, military signals, and so on" as systems of signs. A semiotic literature analyzing the proliferation of signs developed from this idea in Europe in the 1950s through the 1970s and in the United States by the 1980s. Semiotic theory does not try to resolve the philosophical debate between idealists and realists regarding the ultimate reality of the world "out there," but it does attribute to much of culture the same basic structure as language, in particular the signifier/signified duality. Sports, tourism, culinary practices, fashion, photography, architecture, urban design, and even soap powders were studied as sign systems (Barthes 1972, 1979; Gottdiener and Lagopoulos 1986). Places such as the pub, the beach, the home, and the shopping center were also interpreted using semiotic theory (Fiske, Hodge, and Turner 1987). The quest for a semiotic theory of the environment endured less than a decade, however, and its results were not very impressive, in part because it focused on signs to the exclusion of symbols and signals.

At times, we seek a retreat from signs. Situations replete with signals provide this escape. A trip to the beach relocates the body, but, more important, we use it as an excuse to shift from the intense use of signs to a social situation where signs are sparse and signals are much more common. The place of escape is less important than this shift toward signals. In lieu of visiting the beach, we wolf down a good dinner at a restaurant, intoxicate ourselves at a bar, visit a church or temple, have a sexual encounter in the bedroom, take a drive in the countryside. Activities oriented toward shifting our awareness from signs to signals have a special role for modern urban dwellers seeking an escape not just to an alternative place but to an alternative self. Our final *relata,* symbols, also provide an escape from the sign world, but they do so by summing up and concentrating the meaning of signs—standing in for entire texts as a concept, theme, moral, or theory.

Symbols

A symbol for "cat" might be an outline like a silhouette or shadow of a cat. Another symbol for cat might be a gesture, like in American

Sign Language, in which stroking the fingers near the cheek indicates whiskers, implying *cat*. Yet again, we might symbolize cat with a shape like a cat's paw print or a sound like "miaow." These various symbols are entirely different from each other, but all of them relate to cats in a way that is nonarbitrary or "motivated" (Barthes 1967, 50–51). In a symbol, some element of perceived reality is reproduced in simplified form. This means that a symbol for cat used by one culture might potentially be interpreted correctly, through *inference,* by members of a second culture without any teaching or training. In contrast, the word *cat,* a sign, *must* be taught to be understood.

More abstract symbols are also motivated; "our symbol of justice, the scales, could hardly be replaced by a chariot" (Saussure 1983, 68). But the connection between symbol and referent is not always direct. A simplified picture of scales can be used to represent justice, but the image is hardly adequate to indicate all that is implied by the concept of justice. The metaphorical link between justice and balance is what motivates the association, and the visual image of scales, as limited or "overrun" as it is, initiates this link.

Symbols are therefore paradoxically *constructed but not arbitrary.* Symbols have a link, either direct or indirect, to objective reality, although, as many have argued, there is no definitive, complete, or unbiased symbolization of reality. Objective representations occupy the latitude available for artistic representation in a particular culture. In the Western artistic tradition, one takes a particular *viewing angle* when drawing an animal, and one does not attempt to show what is inside the animal (except in veterinary texts). The flow of time is frozen, and the animal is shown in only one position. This is an "outsider" perspective idealized by photographic realism, which constitutes the object as separate from the artist, "out there," clearly fixed in time and space, rendered in light and shade. Native Americans' pictures of animals, in contrast, often do not adopt a fixed perspective, but move over and around the animal, showing all four legs spread out as if the animal were split up the middle, and sometimes showing the animal's insides and disregarding light and shade (see figure 3.1). Though distinctly different from animal symbols in European culture, such drawings are still symbols. Moreover, the moti-

3.1. Animal representations by native North Americans: *(a)* Tsimshian bear, *(b)* Haida shark, and *(c)* Haida duck. From Deregowski 1973, 186. Reprinted by permission of Gerald Duckworth and Co., Ltd.

vated character of both representational conventions means that if given a few moments to study the image, non-Westerners can often "read" Western symbols, and vice versa (Deregowski 1973). The most striking demonstration of the nonarbitrary nature of pictorial symbols is the fact that chimpanzees can achieve more than 85 percent accuracy identifying objects in photographs they have never seen before, using sign language to tell researchers what they see and demonstrating emotional responses such as fear or sexual desire to photographs and drawings just as if their referents were physically present (Fouts 1997, 100–101).[2]

The motivated character of symbols is difficult to appreciate if the chains of association are long and intricate. Some non-Christians find it

2. Whether chimps would be as successful at "reading" non-Western images is not known. Chimps clearly relate to photographs as nonarbitrary representations of real things, as proved by one chimp who would rub photographs of naked men in *Playgirl* magazine with her genitals (Fouts 1997, 153).

difficult to understand how the cross can be a symbol of faith, love, and goodness when the *visually motivated element* is an instrument of torture. For believers, of course, a sequence of signs (the New Testament) clarifies this link between torture and divine love, making the association between torture instrument and spiritual salvation seem natural and essential. A scientific theory such as $E = MC^2$ may require many signs for full explication (the explanatory text), but the theory itself reduces to the equation, which models an underlying reality of the universe. The motivated quality of this symbol is not in its visual form, but in the relationship people have identified between the phenomena it defines; each element of the equation, being a sign, might easily be substituted with another sign, such as $Z = QR^2$, but the equation as a whole (however we might represent the variables and operators) would remain an accurate representation as long as its readers know the conventional meaning of each element, and as long as these elements stand for energy, mass, and the speed of light, respectively. In short, arbitrary *relata* (signs) can be used to build relationships such as equations or other scientific descriptions that are nonarbitrary. The fact that these equations are selective and nonarbitrary makes them symbols, akin to photographs and drawings, even if they are "seen" with the mind's eye rather than with the eye.

Unlike signs, symbols can stand alone without any kind of linear sequence or syntax. That does not mean symbols are free of context. When a national flag is raised and lowered, a nation is instantly associated with the geographical location of the flag pole and, more important, with the institution doing the raising. The flag's purpose is to invoke a rich, complicated, and emotionally charged set of associations between place and polity. Whereas signs help the self to roam through time and space, symbols provide roots. These roots bind actors to stable, powerful, and apparently unchanging aspects of experience. Symbols suggest *relationships* among a wide set of actors, here and there, and in the past, present, and future. They do not permit a person to specify a wide range of particular space-time relationships, like signs embedded in syntax, but instead they imply a cloud of such associations that make far and near, then and now present all at once.

The symbol is, in Barthes's terms, "overrun" (1967, 38) by the wider associations that it evokes: spatiotemporal, social, psychological, moral, and philosophical. Religions and nations are often embodied in symbols because of the powerful emotional link formed between people and these collective identities. People are willing to die for symbols and will react as if their lives are threatened when a treasured symbol is challenged. Symbols of nation and religion are the most apparently overrun in this way, but ideology in general is symbolic.

The space of the symbol is not the systematically structured space of signs, but a deeply emotional space that people discover through their own actions when they find themselves defending something that seems to justify their own existence. When a debate becomes heated, people often try to reduce the problem to terminology (signs) because different sign systems are much more easily reconciled than different symbol systems. Yet for every conflict avoided by clarification of terminology, another conflict is engendered simply because symbols are so often volatile, at least in modern societies without clear-cut standards for the use of symbols. Communicative acts such as burning a flag, hanging a cross upside down, or redrawing a school mascot in a degrading posture are perceived as threatening to country, church, school, society, or community and can arouse intense passions.

A hat-wearing male visitor to the Memorial Student Center at Texas A&M University may be surprised to hear a chorus of "Please take off your hat, SIR." What is *wrong* about the hat in the Memorial Student Center is different in essence from what is wrong when a word is out of place in a sentence. An out-of-place word is confusing or meaningless, whereas a hat on a head in the Memorial Student Center constitutes a kind of blasphemy. Like all blasphemies, this one is a challenge to the authority system that gives the place its meaning. The place is not a shrine, but a multipurpose student center with a bookstore, cafeteria, barber shop, and electronic video games. Nevertheless, it is invested with the metaphysical power to transform a hat into a symbol of disrespect for former "Aggies" (Texas A&M alumni). In the realm of signs, we distinguish between sacred and profane places, but in the realm of symbolism we treat both kinds of places in similar ways, imposing rules on what symbols may go where in relation to other symbols. This power inheres in symbols because they remain within the embrace of a mythical or magical mode of thought in which, as Robert Sack (1980) argues, they are conflated with their referents.

An "out-of-place" symbol does not generate confusion, like an out-of-place sign, but instead arouses passionate responses. It is immediately political. The emotional response to an out-of-place symbol differs markedly from the confusion that ensues when a sign is out of place, a confusion that is cognitive more than emotional. "Good" placements of symbols are, as has been suggested, virtually innumerable (a flag pole can go virtually anywhere), but there exists only a small range of options for the "meaningful" placements of signs in a sentence. The mixed-up sentence "I flag the put up" may confuse people, but it does not *offend* as does a flag hung upside down or sewn onto the seat of one's pants. Note that we often speak of "*the* flag," not "*a* flag," indicating that each instance of the symbol is in place and yet joined magically to every other instance. The idea that one symbol out of place can undermine the state is not so surprising insofar as the state is itself founded on the collective veneration or even worship of symbols (Anderson 1983). Yet the potency of responses to flag burning suggests self-defense, as if people felt themselves to be at some level coterminous with the state, though, unlike France's Louis XIV, they might not have "I am the state" on the tips of their tongues. In short, the symbol is overrun because it means more to us than we can put into words, and it is the external surface of a tendril linking self and Other.

Symbols are all around us: in architecture, clothing, our own bodies, and our vehicles. What distinguishes them from signs is not only their motivated character but also their level of abstraction. As a symbol of aspiration toward spiritual perfection, the church steeple actually reaches upward into the sky. The steeple presumes a mapping of moral space onto four-dimensional time-space, where what is morally superior is literally "on high," and one who breaks commandments is "fallen."

Not all symbols are visual. Consider a kiss (and surprisingly it is not too difficult to consider such a thing analytically if we think in terms of symbols). We are dealing with a kind of touch that is symbolic and as a symbol is differentiated from other symbolic gestures that are less intimate (smiling) or downright aggressive (spitting, biting) by the metaphorical link between softness (sensed haptically) and an affectionate, benign, loving attitude toward something. As a tactile symbol, the kiss carries a tremendous burden of associations whether it is a passionate kiss between lovers, a peck on Grandma's cheek, or a gentle nuzzle of a baby's head. But what is most surprising is that the kiss is not just an event in the here and now (though part of the meaning is at the signal level and is therefore tied to the here and now). At the symbolic level, the kiss reaches out through time, to the past and future: lovers seal their fate with a nuptial kiss, and every kiss thereafter "refers" to that kiss. Symbols are not "natural" in the sense that everyone would choose the same symbol. A kiss on both cheeks reaches out to the community of others who use this gesture, constituting a here and there (the United States is "there" because people in the United States do not normally use this greeting). In a different way, the kiss also creates an intimate place in which the one who is kissing and the one who is kissed have some sort of personal, unique (at least it feels so) relationship. Note that the *context* of the kiss is not only the room in which it is performed, but the culture in which it is understood. The kiss is overrun by all of that which may be revealed of here and there, now and then, through the kiss—feelings only dimly perceived and quite often beyond the range of words, yet similar in weight and veneration to the other symbolic associations we have considered. How little of this is captured in the word *kiss!* The sign and the symbol meet, yet remain strangers.

Landscapes: Between the Sign and the Symbol

In *Social Formation and Symbolic Landscape* (1984), Denis Cosgrove demonstrates that landscapes evolve through time, not just because of physical changes on the land, but because of changes in the construction of meaning. Landscape is "a way in which certain classes of people have signified themselves and their world through their imagined relationship with nature, and through which they have underlined and communicated their own social role and that of others with respect to external nature" (1984, 15). Cosgrove restricts the meaning of *landscape* so as to emphasize a kind of detached and controlling viewpoint generally associated with outsiders, artists, tourists, landowners, and sometimes geographers, whereas "To apply the term *landscape* to their surroundings seems inappropriate to those who occupy and work in a place as insiders" (1984, 19).

From a similar perspective, James Duncan calls landscape "a pervasive and surprisingly disingenuous cultural production" (1990, 3), but his analysis derives from a foundation in signs rather than symbols, and he makes extensive use of the tools that literary theory uses for analyzing signs—metaphor and synecdoche in particular. At the outset of his study of landscape symbolism in Sri Lanka, Duncan argues that landscape "may be integral to both the reproduction *and* contestation of political power" (1990, 3, emphasis in original). Conflict is the essence of landscape in this interpretation, and Duncan argues, for example, that "Neither the production nor reading of landscapes is ever 'innocent.' Both are political in the broadest sense of the term, for they are inextricably bound to the material interests of various classes and positions of power within a society" (1990, 182). Thus, the extravagant building projects of Sri Vikrama, the last king of Kandy (Sri Lanka), were meant to enunciate his authority symbolically, but were challenged, again on symbolic grounds, by rebellious noblemen. Duncan's analysis may be appropriate to Sri Lanka, but as a general principle the idea of a binary logic derived from the study of signs and rhetoric is somewhat problematic when applied to symbols (Soja 1996). A symbol stands alone, without an opposite, although it is embedded in its social context, and rather than search for an opposite and supporters of that opposing signification so as to reveal the conflict behind landscape, we might find it more useful to think in terms of the opportunities a landscape offers its occupants for the performance of self, not always against other selves, but often with and in tacit support of other performances of self (Goffman 1959).

Although both Duncan and Cosgrove have demonstrated in a striking way this symbolic aspect of landscape, it is Tim Cresswell (1996, 2001) who most suggestively *peoples* this symbolic landscape. In general, his studies reveal people as not only the symbolizers *of* landscape, but as embodied symbols *in* the landscape. More specifically, he reveals that people "out of place" in the landscape are not meaningless, like words out of place in a sentence, but instead are destabilizing to society, *oppositional*, like flags or other symbols that are out of place. Simply to occupy space at a particular location is a symbolic gesture. The individual as a body is a part of the landscape and therefore must watch his or her placement in the landscape because the place where he or she is located is always charged with symbolic meaning. To be out of place is not simply to be meaningless or confusing (as it would be if *person* were merely a sign). To violate expectations and occupy space in an unexpected way— to be "out of place"—is at once to challenge the symbolism of personhood, society, and landscape. It is bad or evil (or rather, it symbolizes one of these notions). It shocks, like a piece of evidence that contradicts a cherished myth. People are seldom seen sleeping on the courthouse lawn, dancing in the grocery store, or making love in the laundromat. Although a theft violates a person's claim to property, it does not undermine the concept of property as something of value. Violations of the symbolic construction of reality actually challenge the values ascribed to places and hence are more threatening than crimes. Deviant positionings of the body carry the emotional freight of blasphemy.

The very complexity of landscape as an overrun symbol means that the binary logic implicit in "reproduction and contestation of political power" can become a straightjacket for our thinking. Struggle alone is an insufficient explanation of the entirety of landscape because even as people struggle, they come together to create spaces of shared meaning (Cosgrove 2001). A landscape's elements may be defined in authoritarian fashion, for example by the king who designs a ceremonial reservoir, but people will revise their symbolization of the landscape in response to a reality that is not dictated, in this case if the reservoir turns into a breeding ground for mosquitoes or the dam breaks. This response is a form of environmental adaptation, directly analogous to shedding a heavy coat of fur when the weather warms or migrating to gain access to food sources.

Landscape in the human sense is thus not only a *sign* established by social convention and dominance, but also a *symbol* of what really exists (e.g., floodwater, mosquitoes, heat, hunger) in an objective sense that transcends cultural differences and a *signal* that people respond to in a way that differs in complexity but not in its basic character from other animals' adaptive behaviors.

Acknowledging the Symbolic

Of course, for a symbol to have a social reality, it must be shared, and in the process of sharing symbols (or anything) people constitute communities. A community is a kind of affiliation that carries positive value for people even if they are engaged in conflict with some (or many) other members of the community. Herein lies a second limitation to the conflictual view of landscape developed by Duncan and many other geographers. It is not that the conflicts are less important than the sense of community in a quantitative sense, but rather that they are qualitatively different, political rather than spiritual. Community spirit transcends political conflicts because it mobilizes ritualistic, mythical, or magical dimensions of experience.

It is this element of communication that James Carey wants to indicate in his highly influential *Communication as Culture* (1989). His research emerges out of a different tradition than the works I have previously discussed, one that descends from Max Weber, John Dewey, and George Herbert Mead to Herbert Blumer and Erving Goffman—who largely defined the goals, vocabulary, and interests of *symbolic interactionism*—and to the more focused research agendas of scholars such as Peter Berger. Symbolic interactionism is based on the analysis of social performance rather than social structure, a divide that corresponds to its roots in philosophy and anthropology rather than to economic theory. It also contrasts with the Marxian tradition in its emphasis on the dramaturgical dimension of personal identity construction—the performance of self as ongoing improvisation—which tends to be submerged or denied in Marxian accounts.

This tradition's focus on symbols would seem to indicate that it is grounded in a correspondence to the objectively observable physical world because, as I have argued, symbols are nonarbitrary. The use of *symbol* in the symbolic interactionist tradition, however, is much closer to *sign* as I have outlined it—a kind of meaning based on convention— and this signification provides no epistemological challenge to a binary vision of landscapes based purely on conflict and contestation.

To see the implications of the conflation of sign and symbol we must turn to Carey's interpretation of the news:

> Moreover, news is a historic reality. It is a form of culture invented by a particular class at a particular point of history—in this case by the middle class largely in the eighteenth century. Like any invented cultural form, news both forms and reflects a particular "hunger for experience," a desire to do away with the epic, heroic, and traditional in favor of the unique, original, novel, new—news. This "hunger" itself has a history grounded in the changing style and fortunes of the middle class and as such does not represent a universal taste or necessarily legitimate form of knowledge . . . but an invention in historical time, that like most other human inventions, will dissolve when the class that sponsors it and its possibility of having significance for us evaporates. (1988, 21)

Carey also argues that news "casts" people in particular roles, as American patriots or as supporters or opponents of women's rights, and in doing so it is not always pleasing (one may encounter "bad news"), but it is always "satisfying" in the sense that it confirms a familiar construction of reality. In short, on this account "reality is brought into existence, is produced, by communication" (1988, 25).

Symbolic interactionism highlights the constructed quality of *relata* at the expense of their motivated quality, which results in an expansion of the binarism of the sign from the word to the entire text. The aggregation of signs would seem to produce nothing more than a more complex sign. News is to the world as the word *news* is to the word *world,* a signifier attached to another signifier through convention, a tacit agreement backed by power that *this stands for that*. In contrast, I argue that we must observe and appreciate the way landscapes and news stories are qualitatively different than signs even if they "contain" signs as ingredients. They reveal a form of awareness of the world and may be judged on their inherent correspondence, or lack of correspondence, to the world (a world that can never be known in unmediated fashion, but that nonetheless sets bounds on mediation that are related to practicality). The difference between a news report that exposes the tie between global climate change and human activities and a news report that attributes climate change to natural climatic variability cannot be determined if we adopt a symbolic interactionist or a Marxian and critical theory perspective. It can be determined only if we recognize that communication moves from the space of the sign (arbitrary, conventional, and socially fragmented) to the space of the symbol (motivated, representational, and potentially consensual), a move that occurs as we aggregate signs, which is to say as we consider larger elements of meaning such as news stories or landscapes.

What this literature has failed thus far to elucidate is the difference between signs, symbols, and signals as elements of landscapes. Each of these three *relata* works differently, one at a very concrete level and another at an abstract level. In most cases, the primary analytical frameworks used to understand landscapes, environments, and human attempts (such as news) to understand their landscapes and environments have been derived from studies of sign systems and have therefore included assumptions that are not necessarily appropriate to the study of symbol or signal systems. Such studies posit human reality as freely floating in the space of social relations rather than grounded in various nonhuman constraints. By disregarding natural constraints, these studies contribute to the accelerating breakdown in human-environment relations and inadvertently exacerbate social problems by denying their foundation in material realities.

▲ ▼ ▲

To restate the points made thus far: (1) meaning is composed of heterogeneous constructions; (2) these constructions are made up of various combinations of three basic *relata* with different relations to space, time, material reality, and the communicators; (3) signs are arbitrary, and the "superiority" of one sign over another can be established only through conflict and contestation between human groups; (4) symbols are motivated, and although no symbol is natural or corresponds transparently to reality, some are more quickly proved inadequate by their failure to support adaptive behavior; (5) signals are shared by humans and other animals, and, in the case of unconditional stimuli and conditional stimuli arising from the environment, signals represent a level of reality more directly linked to environmental adaptation than is the reality represented by symbols; (6) texts and landscapes consist of complex arrangements of signs, symbols, and signals and therefore cannot be understood purely in terms of social conflict and contestation because the latter apply to signs only in an unqualified way; (7) texts and landscapes provide people with a basis for engagement with the world that is always shaped and directed by societal conflicts, but that transcends these conflicts because of the signals and symbols constituted at the aggregate level by collections of signs.

The landscapes we inhabit and the places of daily life are not reducible to the social logic of signs, but instead reflect multiple logics. Over time, the logics of the symbol, which is motivated, and of the signal, which can also be motivated, exert an influence on human understanding. Misunderstandings of environments lead to conditions perceived as detrimental to human interests, which in turn support reevaluation of *all three* realms of meaning: symbol, signal, and sign. If people create "bad" places for themselves, then they will eventually refine their communications so as to support the creation of better places. Ideological control or hegemony may slow or obstruct this process, and chapter 4 clarifies its role, but symbols and signals from the environment undermine the lies on which coercion is founded, requiring a shift to the use of force and the concomitant erosion of ideological legitimacy. Social structure or the terrain of conflict is not the ultimate ground for all meaning.

Regarding the implications for the extensible individual, each of the three *relata* sets people in relation to the world in a particular way. It scarcely helps our understanding to apply simplistic labels such as *direct* and *indirect* or *abstract* and *concrete,* but in general we should recognize that the self is connected to other selves and to environments both directly and indirectly—through signals, symbols, and signs. With every articulation or enunciation, several strands of connection to the human and natural world are implied. Although Deleuze and Guattari (1983) promoted the term *rhizome* for social relations, the necessary corollary of such a weblike social space is not only the decentered and fragmented self envisioned by postmodern social theorists, but at the same time (on another level of analysis) the extensible self outlined in the previous chapters. The extensible self is not the occupant, but rather the substance that makes up the rhizomes of communication, and some of these rhizomes are rooted in the nonhuman world, in "nature," if we understand by that an entity that lies beyond our representations and is only partially captured by any given symbolization. In place of a rhizome, we may think of an antenna or a tendril. Perhaps most evocative is the metaphor of the pseudopod, a temporary appendage extended by a single-cell organism to engulf its food (Adams 1995). Extensibility is material and is an encounter with the not-self, assisted by *relata,* so as to overcome the boundaries of the self.

Movement through physical and virtual spaces is simultaneously a means of extending the self into society and nature and an ongoing encounter with the extensions of others. Let us look more closely at the physical expression of this encounter and see how it is grounded in the landscape.

A Walk in the City

You step out your door to take a walk in the city. The scene is rather monotonous: house, house, house, house, street . . . house, house, house, house. Like many Americans, you have to walk more than five minutes before the scene changes substantially. There are no stores here, no craft shops, doctor's offices, or restaurants. Each house is different, yet they all stay within rather narrow bounds of appearance, as do the yards. The houses are one story high, painted in muted colors, roofed with asphalt shingles. Most of the yards have one or two large trees, a row of bushes directly below the windows, and a clump of flowers next to the mailbox or beside the driveway. The rest, it goes without saying, is grass. Grass care does vary from yard to yard. Some yards are short, springy, uniform green turf, rising straight from the edge of the driveway and the curb. Other yards are a yellow-brown mixture of clover, thistles, and grasses that sprawl out in tendrils across the curb. Walking past, you sense that the greener, neater yards are better cared for.

This perception of "care" is a social construction, an inadvertent "reading" of a symbol, and this reading has its roots in a Western (Christian and capitalist) distinction between self and nature (Hough 1995).[3] A "cared for" lawn is an expression of a particular philosophy that stands apart from nature and reworks its elements in the name of care and improvement (Howarth 2001). Of course, grass does not ask to be taken care of; a "healthy" lawn does not wake up in the morning and count its blessings. To care for a lawn is, in essence, to intervene in and redirect a natural process as a way of demonstrating one's place in the social world (Bormann, Balmori, and Geballe 1993). This demonstration leads to an ironic situation: the "cared for" lawn is kept green and weed free with chemical herbicides and fertilizers, which harm birds, fish, frogs, and other animals, both directly (when mistaken for food) and indirectly (when carried to rivers and lakes by runoff). As you appreciate the "cared for" yard, you participate in the exchange of a symbol—tidy green lawn as a symbol of care. "Lawn" as a symbol is "overrun" by its mapping onto a model of care that specifically works against what Michael Hough (1995) calls the "natural process," an unimpeded (though not necessarily pristine) interaction of plants, animals, climate, and other nonhuman forces. An alternative symbol of the lawn based on a scientific model of the ecosystem might be implemented. This model might define "care" as allowing the lawn to revert to a forest

3. For more on the suburban landscape as "discourse materialized," see Schein 1997.

or meadow ecosystem, perhaps even a wetland. Any such model is not only rare but *deviant;* most American cities levy fines on property owners who try to apply this model. It is perhaps easier for the majority of people to feel affection when they can control something and twist it to meet their moral or aesthetic whims (Tuan 1984). This link between dominance and affection therefore becomes the societal norm guiding lawn care, and laws are written to discipline those who might prefer a self-regulated ecosystem outside their door.

The symbolism of care is complex. It points to the embodied agent. What is deemed "good" and "healthy" has an odd link to personal hygiene. Neatly trimmed hair, brushed teeth, and an ironed collar are testimony that their owner has his or her life under control. A lawn that looks trimmed, brushed, and ironed works exactly the same way to symbolize the physical, mental, and emotional self, to *extend* that self into the material world, with effects that are scientifically understood but beyond the scope of public discussion. Because of this link to personal hygiene, it is not surprising that the standard complaint lodged against an "overgrown" lawn is that it constitutes a "health risk" (Bormann, Balmori, and Geballe 1993, 33). The irony is that exposure to pesticides appears to be one of the more ominous health risks people face today. The symbolism of hygiene leads to an unhealthy environment, a kind of systematic destruction under the guise of care. Each yard is a self extended into plant life so as to join the silent consensus that what is unnatural is necessarily cared for and vice versa.

Yet into this orderly system a further ironic twist is added by the cats that step out forthrightly to meet you, meowing and rubbing against your legs while their owners remain aloof. Nature seems to be extending itself into this environment as well, suggesting that person A might run his fingers over the extended body of person B, with a rather shocking sort of vicarious promiscuity. Animals are uncivilized, and we love them for it, or, more precisely, for the inhuman, unruly signals they send to any (humans or animals) who may be paying attention.

As you continue walking past floral and faunal extensions of human selves, you also pass cars, pick-up trucks, minivans, and SUVs. These machines are not mere utilitarian objects for which meaning and function are identical. Instead, much like hairstyles and grass styles, the vehicles are extensions of selves. On the bumper of a Chevy Tahoe Z71 four-by-four is a BUSH/CHENEY 2000 sticker and a National Rifle Association sticker. Across the street is a boxy fifteen-year-old Volvo with a WAR IN IRAQ? NO sticker, sharing a driveway with a dented Toyota with an "Animal Friendly" license plate and a NO MORE SLAUGHTER: SPAY/NEUTER YOUR ANIMAL COMPANIONS bumper sticker. You pass more parked cars: an Acura discretely sporting a small "Jesus fish" (a silver metal outline of a fish inscribed with "ΙΧΘΥΣ"); a Jeep Grand Cherokee with a white soccer ball decal in the back window surrounded by the words "Patty" and "Lakeridge Soccer"; and a Subaru with three decals in the rear window: SUNY-Potsdam, ND Alumni, and the Texas Longhorn symbol. You can infer political stances, hobbies, religious beliefs, sports interests, income groups, academic credentials, and consumption patterns of the owners of these vehicles from the decorations on their windows and bumpers, which combine symbolically with the vehicle make and model to extend the self into the material world.

This is not a sign system; no rule says that a BUSH/CHENEY sticker must not go on a Subaru or that families with soccer-playing children have to drive SUVs. Each element is flexible yet emotionally laden, like a cross or a flag. The vehicles are overrun by meanings, not just commodity fetishes (signs of wealth), but another silent conversation (at the symbol and signal levels). These extensions of self show less consensus than the lawns. By "extensions of the self," I mean not just that the wheel is an extension of the foot, as Marshall McLuhan argued (1967, 31–33), but that the vehicle one drives becomes a metaphor for the self—a mechanical body, a cyborg technology (Haraway 1985).

The car even more than the yard stands for the *public* self. As such, it is a perfect illustration of Bourdieu's notion of *habitus*. In Bourdieu's condensation, "Social subjects, classified by their classifications, distinguish themselves by the distinctions they make" (1984, 6). *Habitus* is a learned pattern of preferences in music, art, literature, architecture, fashion, and so on—an interpretative framework that positions one in society so one can position others. To place a sticker or decal on one's car is to add a mark of identity to one's most publicly visible possession and therefore to add *distinction* to the vehicle, a mark of good taste (for those in one's chosen community), an endorsement of one value scale, and a rejection of other value scales.

If a vehicle is a billboard for a particular lifestyle, level of consumption, value system, or segment of the collective memory, like the lawn it offers no channel for response. The street, which used to be a place of debate and discussion in the "walking city" of the nineteenth century, is now a "machine space" where debate is silenced: allusions to the self, glimpsed in motion, evoke competing values and symbols of status, but there is no place in which to weave these threads (Horvath 1974). The driver with the Jesus fish can only dream of converting the driver of the "Darwin fish," and the latter of winning a humanist convert. In this dance of deferred debates, one is proud to drive a Chevy Tahoe with a BUSH/CHENEY sticker because the tall and powerful SUV confirms the driver's belief that this is a powerful (and therefore good) pair of leaders; the motif "power equals virtue" dominates. The Honda with an "Animal Friendly" license plate reveals an ethos of moderate consumption (high gas mileage) and humility (small and low vehicle) matched with pro-animal sentiments (the sticker). The dominant ethos is "moderation equals virtue." The two drivers silently shout their values, but deliberation or rapprochement of perspectives is impossible.

Various forums of public discussion and dialogue in cities of the past, as flawed as they were, are increasingly being replaced by a kind of urban-scale aquarium in which vehicular cyborgs swim about, their communication consisting of the silent display of symbolic decoration with no thought of receiving a reply. The social status associated with vehicles such as Grand Cherokees, Chevy Tahoes, and Humvees adds "weight" to the symbols they sport on their bumpers, while these same bumpers cause disproportionate deaths among the drivers of physically lighter vehicles such as Honda and Toyota sedans.

David Brooks's book on bourgeois bohemians, or "Bobos," lampoons the passion for bigger, faster, stronger equipment, which is not reflective of the environment in which Bobos live:

One of the results of this trend is that there is an adventure gap opening up between members of the educated class and their belongings. The things they own were designed for more dangerous activities than any they actually perform. The hiking boots that were designed for the Andes spend most of their time in the farmer's market. The top-of-the-line fleece outer garments are used for nothing more strenuous than traversing the refrigerated aisle in the Safeway. The four-wheel-drive vehicles are never asked to perform any ordeal more treacherous than a bumpy road in the slush. But just as in the age of gentility hypocrisy was vice paying homage to virtue, so today among the Bobos rugged gear is comfort paying homage to adventure. (2000, 90)

Skirting this silent symbolic exchange among ever-larger and more powerful vehicles, the pedestrian is defined simply by his or her lack of a vehicle, a crucial absence that becomes in itself a symbol. Though we speak of the "homeless," what makes these unfortunate people stand out is that they are carless (so they carry dirty bags or push wire carts around the city in a ritual of public humiliation). Two of the most visible groups of the carless are children and the very old, groups marked by associations of disruptiveness (each is uncivilized—which is to say each communicates as often in signals as in signs), coupled with an economically threatening tendency to be dependent.[4] The same associations are attached to the homeless.

The bicycle is contaminated by these associations—made into a symbol of immaturity or failure. Bicycle riders often feel a need to recuperate their vehicular selves from this infantilization by adopting the symbolism of nature's power, coupled with the symbolism of power over nature. So all bicycles, even those used in the city, have become "mountain bikes" and are designed with heavy-duty tires and spring-

4. In College Station, Texas, the home of Texas A&M University, I lived only twenty minutes by foot from my office and consequently decided to walk to work. This turned out to be difficult because once a week or so someone would pull up alongside me and ask if I needed a ride. Carlessness was so out of the ordinary in this environment that it was assumed to indicate some kind of problem.

loaded frames for use on rocky trails, although these features add weight and make them slower on paved roads where they usually are ridden. The functional requirements of urban transportation are of little consequence in the design of this transportation precisely because people perceive their vehicles in terms of personal identity more than in terms of environmental appropriateness. We are operating here at the signal level, and it would be foolish to situate the origin of attitudes about vehicles simply in social structure and therefore at the level of sign systems. As weak and naked primates, humans have always depended on their tools to establish dominance within and between groups, regardless of social structure. Now such behavior is both anachronistic and dangerous because it drives excess consumption, which exacerbates ecological instability, but it is not a unique result of capitalism and its fetishistic sign systems.

When the sidewalk disappears for a half-block, you must walk along the side of the road between parked cars and moving cars. The signal you receive (it is a signal because it would be understood by any mildly intelligent animal) is one of vulnerability. This signal reinforces the symbolism of carlessness just discussed. Every passing car is a signal to be alert, but your elevated attention becomes itself a symbol of powerlessness rather than a basis for action because an out-of-control vehicle is actually impossible to escape. In short, the lack of a sidewalk in this part of town signals vulnerability and symbolizes the walker's failure.

Mingling with the parallelism of signal and symbol and their construction of walking as marginal is the fact that as you walk, the back of your neck becomes damp from perspiration. If this were happening in your car, you could turn on the air conditioner, which not only cools your body, but also conveys a sense of power—the primal power of being in control of one's environment. Conversely, that which must simply be tolerated sends a signal that one is powerless, which increases the level of the discomfort itself by lending a meaning of "weak self" or, better yet, "small self," self with minimal extensibility. In this way, technological appropriation shifts simple meanings such as "hot" and "cold" from the level of physiological signals to social symbols.

There is a profound irony about this multidimensional link between walking and powerlessness that society has constructed. The walking body is, among other things, getting exercise that helps it maintain or increase its physical strength and stamina. The most dangerous thing about vehicular transportation is not that it ends in a fatal crash in the rare instance, but that it dooms city dwellers to high rates of heart disease and obesity. With Baudrillard's procession of simulacra in mind, we might see the vehicle and the road system together as something that "masks the *absence* of a basic reality," that reality's being personal physical power and its decline through immobility. Extension of self into machine as a second body, a *cyborg* self, leads to an illusion of power, but at the same time causes the atrophy of the organic body's power (unless of course the original body is periodically inserted into another kind of machine at a recreation center and exercised). The body thus kept in working order through pointless labor is symbolically very different from a body that serves as a vehicle; it is more like a pampered pet in a cage than a direct extension of the self; it is something that "masks the absence of a basic reality" and is one step away from hyperreality, in which it bears "no relation to any reality whatever" (Baudrillard 1983, 11). The cyborg self denies nature both inside (the body, the extended self) and outside (the city, the ecosystem).

At eight and one-half minutes into your walk, you approach a major thoroughfare. You begin to detect signals of its proximity: traffic sounds and the summery odor of exhaust mixed with french fries. This road bounds your neighborhood on one side. As in most American cities, a major road zoned for retail and light industry makes a distinct edge between different districts of the city, dividing neighborhoods, schools, and ways of life. Kevin Lynch (1960) points out that such edges are fundamental elements of people's mental image of the cities in which they live, along with landmarks, nodes, paths, and districts. Some edges lie across urban space like a great rift between parts of the city, others act as a "uniting seam" bringing the parts of the city together. Edges are often also paths, ways to get from one part of town to another.

The street you turn onto now constitutes both types of edge: a uniting seam for the mechanically extended cyborgs that roll about the urban landscape, but a divide for the handful of pedestrians who scurry in the gutters. It is a primary thoroughfare for vehicles moving north and south, but a path for only the most stubborn or desperate pedestrians. Crossing it takes patience, and walking along it means you must watch your step because the sidewalk dips frequently to allow access to parking lots. The concrete is broken and uneven, littered, strewn with gravel, and overhung by signs. The symbolic language of the space is designed for vehicles: the white lines between parking places, the inconvenient rise and fall of curbs, the arrows separating turn-ins from exits, the broken sidewalk and the insistent roar of vehicles—all send signals that you are out of place here if you are on foot.

In another time and place, the walker was the quintessential urbanite. Nineteenth-century Paris generated the ideal of the *flâneur,* an urban stroller whose primary enjoyment was looking at the urban scene. The ideal *flâneur* was, not coincidentally, a man; women were expected not to move about the city or to do so only in a goal-directed way, as when going to work or on shopping trips (McDowell 1999, 152–56). Elizabeth Wilson argues that although male, the *flâneur* was a feminized identity in light of his interest in observing and appearing rather than participating and because of his passion for consumption (see also Wilson 1991 and Wolff 1985). Still, the *flâneur* was a male body interested (often to a disruptive degree) in female bodies, and the female bodies who made themselves at home on the streets were socially constructed as sexual objects, as "streetwalkers" (Solnit 2000). As machine space has replaced the city of the *flâneur,* the *flâneur*'s fluidity has given way to the solidity of three-ton "off-road" vehicles and to the city roads and parking lots dominated by them. This aggressively masculine form of embodiment that has shaped American cities is a combination of person and machine, a cyborg with its desires sublimated to the nostalgic conquest of nature: from contact to display, from communication via speech and touch to the silent ballet of cold metal.

Where the qualities of place have not been entirely sacrificed to vehicular access, a walk can of course be a sense-fulfilling experience unlike any other. Even in the most sterile urban landscapes, dawn and dusk are still sweet, especially on summer weekends with the scent of backyard barbecues and on winter days when a fresh snow covers the pavement. A cat rolling on the sidewalk or a squirrel peeking around a tree trunk ironically has the ability to humanize the landscape because they make it feel lived in even when the human inhabitants come and go encased in metal, via remote-controlled garage doors.

If one should be so lucky as to live where the terrain is not entirely flat, the resistance offered to the muscles while climbing uphill, the breezes that play at the top of a hill, the swinging gait of descending steps are still available to the walker—faint signals of nature's topography. Although the sounds of conversation drifting from open windows are growing rarer with the diffusion of air conditioning, one can still sometimes catch a fugitive stream of signs as it passes from private to public space, symbolizing a "community" that is locally based and comfortable. These experiences lack the hard-edged "extreme" sensory qualities of machine-enhanced recreation, and to write of them even in bland and prosaic words seems romantic because they are alien to the cyborg's power-oriented sensibility. A love of walking was popularized, after all, by some of the most famous romantics: William and Dorothy Wordsworth, Henry David Thoreau, and Walt Whitman. It requires a truly out-of-place and out-of-period attitude to defend promenades, strolls, hikes, and other embodied engagements with place in the "information age." But we must recognize the sign bias in this attitude and revalorize such romantic pastimes as a space for the reassertion of marvelously uncivilized signals.

You pass a bus stop where seven people are waiting around a single concrete bench: a black woman and her young boy, three workmen in checked shirts conversing in Spanish, a white woman with dirty hair who is chain smoking, and a blind man with a cane. You glance at these people as you pass. The workmen are sweaty, the mother is scolding her child, and the smoking woman coughs. Eyes meet yours, and the confrontation of *habitus* results in reciprocal judgments. Every group carries a symbolic significance to your eyes. Of course, you, too, are a symbol for each of them. Male or female, old or young, black or white, your appearance as you pass means something to them just as their appearance means something to you; it says something to them about the place. Moving through public space, you are inevitably what Calvino calls in another context "an emblem among emblems" (1974, 23).

Your sense of this place and its occupants is not one you care to savor. It would take many words to capture and convey this symbolism of place, but it stands in marked contrast to the coffee shop where you buy a latté most afternoons or your favorite restaurant that you go to every week or even the grocery store with its air of plenty. In most American cities, the bus stop is a place for people at the margins: recent immigrants, the poor, people who have lost their driver's licenses because of substance abuse, teenagers, the physically disabled. Time at the bus stop moves slowly because buses are infrequent, and there is nothing to do. The tedium, the lack of space on the bench, the exposure to the elements, the socially marginal company, the unpredictable arrival of the bus—all these factors send the bus rider a symbol message that he or she could and should find a better way to get around. This message resonates in other signals and symbols of the place: a paper cup on the pavement that twirls every time a vehicle passes, water that sprays from the wheels of passing cars, music that throbs from a car window then fades into the dull traffic noise. To be in this place is to partake of its sense of being uncared for, a sense as prominent as the cared-for quality of the manicured lawns you passed earlier. As a public place, it reflects no one extensible self in particular, but adheres to everyone that passes by or through it, a kind of unwelcome extensibility that people try to avoid, yet cannot because this place is a node, a place in which lifepaths converge.

The symbolism of "care" and control encourages us to retreat from multisensory engagement with the city. Tuan argues, "An object or place achieves concrete reality when our experience of it is total, that is, through all the senses as well as with the active and reflective mind" (1977, 18). The bus stop is real, but it is not very inviting. Much more inviting is the air-conditioned interior of your car from which the city appears as a stream of luminous images: silent, odorless, and softened by movement. By Tuan's criteria, the city surveyed through the windshield is not quite real, but this version of the city is a delightful one: shining, fluid, at your command, constantly opening before you, yielding and pliant. As Doreen Massey reminds us, "spaces and places, and our sense of them (and such related things as our degrees of mobility) are gendered through and through" (1994, 186). For the cyborg merged with an SUV, four-by-four, Harley, or Humvee, the vehicle is a masculine body fantasy—huge, hard, packed with force, and the city itself is an ideal conquest: a smooth unfolding interior space that asks no one *to* return and demands nothing *in* return.

You pick your way along the sidewalk on this thoroughfare, focusing on the signs overhead. These signs are an odd mix of the ubiquitous and the unique: Austin Guitar School, Goodwill, Dan's Hamburgers, McDonald's, El Caribe Mexican Food, Blockbuster Video, Seoul Asian Food and Gifts, U-Haul. Who could not recall the color of the McDonald's arches or label the main items on the familiar backlit menu? Almost equally ubiquitous in the United States are Blockbuster's yellow hexagonal letters on the familiar blue awning, or the U-Haul sign with its black square letters on five separate white and orange squares. These "signs" are of course signals. They say (you, here) "consume" (now) because advertisement has associated them with sensory pleasure, or they at least symbolize a predictable way of doing business. The most familiar signs are more likely to be remembered and sought out later—landmarks in Lynch's terminology. Landmarks are signals and symbols at the same time. As symbols, they evoke an array of memories sedimented in time (assuming one has visited such places many times) and layered in the six senses. As a place, McDonald's is real, but distributed here and there through space and time; it is like a parallel universe one enters through thousands of scattered doors. The less-familiar signs on this strip may be odd enough to catch a driver's attention, but they lack the reassuring symbolic link to past experiences as well as the Pavlovian association with pleasure and security.

You have reached your destination, a Blockbuster Video store, where you release a video cassette from your sweaty hand and resist the impulse to rent another movie right away. This is the mission that called you forth into the city, and it stands now like a riddle. How does the modern city reflect the communication technologies such as videotapes that have silently infiltrated its cracks and pores? The exuberant predictions of the demise of cities and the rise of the "electronic cottage" have not yet come true, though many homes are now thoroughly wired (Toffler 1980).

The changes that have taken place have more to do with social networks than with spatial patterns. Although more than half of Americans know all of their neighbors, these ties are not likely to be strong.[5] Although it is easy to lament the erosion of place-based community, the ease of vehicular access makes possible the maintenance of ties of friendship and family across distances that would have seriously eroded these relationships in the past. Most likely the video was recommended by a friend, colleague, or relative, and tomorrow's discussion with this person, whether face to face or across thousands of miles via the Internet or cellular or traditional telephone, will move through the virtual space of the video: "Why did she leave him?" "How did they get the dog to open the refrigerator?" "What other films was he in?" and so on. Like a physical gathering place, a film or television program provides a common experience for millions of persons; it gives them a particular perspective on the world that they then can judge among themselves. This perspective may not always be profound, but it resonates with shared cultural values and fulfills many of the roles once fulfilled by the most sacred places (Adams 1992). One is free to enter or not enter this "gathering place," free to enjoy or hate what one experiences there, and above all free to interpret the experience according to a range of different decoding schemes (Fiske 1987). Nonetheless, the space of television parallels the physical space of the city, and insofar as people have limited free time, it competes with time in which people might be engaging with others face to face and constructing a more interactive public sphere. Perhaps the clearest impact of media on the city is in the siphoning off of leisure time from city spaces, which results in an impoverishment of those spaces, particularly for senses other than sight.

As you head home from the video store, you decide to take a differ-

---

5. Data are from Wolcott 2002. The percentage of respondents who knew all of their neighbors was 58; the percentage who had never sat down to eat with any of their neighbors was 45.

ent route, a shortcut through the parking lot of a state office complex. A moment later something moves in your peripheral vision. The quick movement startles you (an unconditional visual signal), and your body responds with a jolt of adrenalin. Turning, you see it is only an electric golf cart approaching you; the combination of speed and silence caused the panic signal. The driver, a uniformed security guard (new startle signal, conditional), says, "Excuse me, sir, can I help you find something?" His presence in the cart, his uniform, and his manner indicate that the question is in fact not a question, but rather an expression of territorial control. "No thanks" is not an acceptable answer.

Human confrontations are often resolved through words, but not necessarily through words *as signs*. It would be grossly underestimating the words you have just heard to interpret them as signs. The guard's physical presence, position, appearance, attitude, and gestures "say" more than the words in the language of symbols and signals. Your presence has been identified by a representative of authority, and you are out of place, an outsider inside the territorial boundaries of "public" property. To the guard, your bodily presence in this place is a symbol of disorder and a signal that order must be restored. "Public security" is defined (in laws and regulations) such that the public must not occupy public space unless they have "business" there, or at least appear to, which means arriving in the "proper" clothing, during business hours, driving a car.

Writing in the late eighteenth century, Carl Moritz complained of England's provincialism: "A traveler on foot in this country seems to be considered as a sort of wild man, or an out-of-the-way-being, who is stared at, pitied, suspected, and shunned by everybody that meets him" (quoted in Solnit 2000, 83). Although English attitudes became more favorable toward walking in the nineteenth century and its associations with irrationality and lawlessness declined, the walker remained a socially marginal figure associated with poverty, alienation, and resistance to progress both in England and the United States (Adams 2001). A century of peripatetic (walking) literature romanticized the walker as one who thinks and perceives differently, positioning himself or herself outside of the flow of history. But romanticism had little hold on twentieth-century America, especially with the rise of automobile dependency.

From the nineteenth century onward, many changes seemed to be dictated by technology itself, including an ever-accelerating movement. In fact, a major part of the change was a positive feedback loop in cultural symbolism: urbanization and the rise of the market economy eroded attachment to place-based communities and normalized an urban lifestyle in which distanciated communication is increasingly substituted for face-to-face communication. All of this reduced the link between foot transportation and community participation, a link further eroded by the increased threat and inconvenience of walking in machine-dominated spaces. The decreasing probability of reaching useful destinations on foot compounded the symbolic construction of walking as useless. As people walked less, they came to depend on vision as the primary way of knowing their surroundings—the only sensory mode not shut off by the bubble of the automobile.

The security guard's response is one that William and Dorothy Wordsworth, Henry David Thoreau, and Robert Frost would have recognized. From the viewpoint of authorities who control territories, a walker is altogether too free to be condoned; his or her mobile presence symbolizes a world where territories might blur together in a messy way and territorial authority might be compromised (Cresswell 1997). One of the prime mechanisms for exerting control over people, objects, and processes is the imposition of territorial boundaries (Sack 1986). People moving under their own power pose a threat because they are too mobile, too slippery, too capable of scaling walls, wriggling through gaps, picking things up, and hurrying away undetected. Vagrancy laws and trespassing laws were written to try to control this threat, and by the late nineteenth century a new category of criminal was created, the tramp, someone who traveled from place to place and did not work. Tramps were subject to punishment in the form of forced labor, which effectively transformed them from a threat into a resource (Cresswell 2001).

Retreating from public territory to circulation space—the street beside the parking lot—you once again pass by houses. You are now in place, and before long you are home.

▲ ▼ ▲

In the spirit of a coda (a passage of music formally ending a composition), let us pause to consider the phrases "I am home," "I come home," and "I go home." The grammar of these phrases reflects the special ontological status of home. We cannot substitute "I am work" for "I am at work" or "I go school" for "I go to school." The same elision of the preposition is true in the second and third person. To say "I am home," "I came home," "I go home," or even "I like being home" is to convey the idea of being-in-place without a preposition. The fact that the language is "designed" this way suggests a condition of fusion or merging between person and place. Language in effect confirms that in terms of place and extensibility "there's no place like home." Home is an extension of self par excellence.

This book began with a quote from Tuan, "Humans are language animals" (1991, 694). We can now see that what this means is rather complicated. Like other animals, humans use signals, and signals constitute for people a set of responses to actions and environmental conditions difficult to control fully with the conscious, rational mind, a situation that indicates our kinship to animals as well as the fact that communication is not uniquely human. Furthermore, we can see that signals are essential to animals' adaptive behavior, suggesting that at least signal communication in humans is not a closed circuit within a purely social (economic or political) space, but rather a conduit between people and their environments that predates humanity itself.

Most of the time, however, people mix signals with signs and symbols, two other *relata* that are essentially human in character. Although a wide range of animals, including dogs, cats, and horses, can learn to respond to human signs, the signs *when used as commands* take on the characteristics of signals. Some animals, however, can use signs, most strikingly our nearest relatives, the chimpanzees, so we must avoid making overly anthropocentric claims about our communicative talents. We also should remember that our "evolution" sometimes results in a simplification of communication to patterns reminiscent of more "primitive" animals. Vehicles on a road resemble fish in a river to the degree that their interactions depend on the exchange of signals rather than signs. Theories that "explain" communication in terms of, for example, political economy are missing a vital connection to environment that grounds communication in the material world. The irony is that such approaches are often considered "materialist," whereas their essence is closer to philosophical idealism or symbolic interactionism.

A grounded view of communication can recognize that, unlike animals, humans extend their bodies imaginatively into animate and inanimate objects, even while using forms of communication that were developed by animals through evolutionary processes. To be an extensible individual is therefore to inhabit a space defined by a complex system of signs, symbols, and signals, as well as a physical space. To be an extensible individual is to know one's place through the disembodied media of signs and symbols as well as through the embodied media of signals. It is to be a creature of heterogeneous logics.

Although signs, symbols, and signals constitute the *content* of our communications, we cannot understand them without shifting our attention to factors that guide the flow of information, the exchange of symbols, the geography of interaction, and the access of persons to other persons or groups. These factors are, in effect, heterogeneous spaces: social relations, technological systems, and physical environments.

The next chapter takes up these vital contextual issues.

4

Communication Context
Parameters of the Self

> [M]edia, like physical places, include and exclude participants.
> Media, like walls and windows, can hide and they can reveal. Media
> can create a sense of sharing and belonging or a feeling of exclu-
> sion and isolation. Media can reinforce a "them *vs.* us" feeling or
> they can undermine it.
>
> —Joshua Meyrowitz, *No Sense of Place*

IN THEIR 1944 CLASSIC *The Dialectic of Enlightenment,* Max Horkheimer and Theodor Adorno complained of a "devalued language" that
stifled social reform at its origins, at the very moment of articulating
ideas. They argued that Enlightenment thought, in striving to replace
mythical thought, had adopted instrumental rationality—economically
productive, control-oriented ways of knowing—as its paradigm. Instrumental rationality has all the characteristics of a mythology within
modern culture, defended passionately and ritualistically. The taken-for-
grantedness of this type of knowledge shapes the language available to
those who seek to understand or change society. Its hegemony makes
the communication of an alternative, noninstrumental social agenda
difficult or impossible, regardless of the signs, symbols, or signals one
manipulates ([1944] 1972, xii).

This social science perspective arising from interwar Germany rejected part of the Enlightenment model (not surprisingly, considering
their social context), specifically the idea of progress achieved through

instrumental rationality. "There is no longer any available form of linguistic expression which has not tended toward accommodation to dominant currents of thought; and what a devalued language does not do automatically is proficiently executed by societal mechanisms" ([1944] 1972, xii). The authors saw instrumental rationality as a symptom of a fundamental malady of Western society. "It is characteristic of the sickness that even the best-intentioned reformer who uses an impoverished and debased language to recommend renewal, by his adoption of the insidious mode of categorization and the bad philosophy it conceals, strengthens the very power of the established order he is trying to break" (xiv). What is wrong with the language and the "mode of categorization" cannot be solved within language—for example, through better grammar training (knowledge of signs) or through the logical formulation of statements (better symbolization). Communication's illness reaches beyond content to encompass the entire social structure, in particular its dependency on *domination of one human group by another*. This concern raises fundamental questions about the limits imposed by context on the latitude of signs, symbols, and signals. To address these questions we must first define what context includes:

1. *The specific social relations between senders and receivers of communication,* including expectations about the communication act as it relates to participants' roles in the communication act;

2. *The physical and technological arrangements linking senders to receivers,* including both networks (such as the Internet) and containers (such as rooms) that give a space for communication;

3. *The structure of society,* which includes economic, legal, bureaucratic, and political modes of organization at various geographical scales that are experienced as communication contexts "nested" within one another from the home to the tribe, community, nation, and world;

4. *The general character of human-environment interactions* in a region or regions where the communication occurs.

These concerns are hallmarks of an interpretation of communication that Judith Stamps (1995) calls "materialist." Her broadly defined materialism includes not only Marxist and Marxian approaches, but also the thought of Thomas Hobbes, John Locke, Epicurus, and Galileo.

These theorists base their understanding of the world on material reality as revealed by sensory experience rather than on the abstract thought forms of the Platonic tradition (1995, 16). Their approach is inductive rather than deductive.

Communication research of this type also foregrounds context and can be called "contextualist." It perfectly complements the content-based approach outlined in the previous chapter, which proceeds deductively from the basic elements *(relata)* of content. It does, however, articulate with that approach insofar as the meanings given to signs depend on a particular mode of reasoning applied to the signs, and a range of such modes of reasoning is made available as a communicational resource by the society in which one lives. "My choice of reasoning mode is therefore analogous to the craftman's [*sic*] choice of trade tools; in both cases, I am influenced by my subject matter, by the experiences I have accumulated in the past, by my present milieu, and by my hopes and fears for the future" (Olsson 1975, 52). Although words depend on a mode of reasoning, and this mode in turn depends on elements both of content (subject matter) and of context (present milieu), we can still separate content and context issues—the medium (broadly defined) from the message (narrowly defined). This distinction allows us to proceed as if the line between content and context were clear-cut (at least for analytical purposes).

The room in which one speaks is a bounded communication context with a physical manifestation. Walls define an inside and an outside, limiting the size of the gathering and (in conjunction with other, larger containers such as buildings) determining who can be part of the communication situation. The room, the building, the city, and the nation are nested contexts. All such physical containers are products of the material relations of production and consumption, which also create insides and outsides: investors speak mainly to other investors, pipe welders speak mainly to other pipe welders, and so on because they share physical places. Furthermore, in a single room constructed by a single social group, one may show a film, hand out printed material, give a PowerPoint presentation, and so on. These media of communication (celluloid, paper, computers, projectors, etc.) transmit ideas only to those who understand them and are available only to those who know how to construct or manipulate them. Therefore, such media again define an in-group (those who are literate in a given medium and have access to it) and an out-group (who are illiterate in that medium or otherwise denied access to it). Therefore, places and media work in more or less the same way to define who is in and who is out of a particular communication situation.

This recognition is the essence of the contextual perspective, and it forces us to recognize that we cannot focus on communication merely as a *language,* even when we make the necessary adjustment to thinking in terms of signs, symbols, and signals. Communication is *action,* and like all actions it takes place in a world of physical, social, and technological bounding structures. Content cannot stand alone, but always moves through and across structured contexts: geographical and architectural spaces, social institutions, and technological devices.

In this chapter, I trace the development of one strain of contextual media theory, the chain of ideas leading from Marxian historical materialism to Habermasian communicative action, emphasizing social relations as communication context. I examine views that derive all axioms from context—that is, contextualist positions. I then apply that body of thought to understand a "trip" on the Internet.

Marx and the Context of Communication

Karl Marx's materialist vision of communication is sketchy, but we can see in it the connections between two kinds of communication contexts: social structure and geographical space. Marx understood communication largely as the means by which capital is able to achieve "the annihilation of space by time."[1] Whether on the factory floor, in the train yards, or in the layout of the city, costs of distance were overcome by the institutionalization of a work regime based on mathematically

1. This phrase, often quoted by David Harvey and other economic geographers, is unfortunate because time, as an impediment, is similarly annihilated. If both space and time are annihilated, the annihilation of space is not "by time," but by the diffusion and adoption of technology.

measured and compartmentalized time. We might equally well speak of "the annihilation of time by space" because, according to Marx, time broken down by the organized production regime into segments of finite duration "sheds its qualitative, variable, flowing nature; it freezes into an exactly delimited, quantifiable continuum filled with quantifiable things . . . in short, it becomes space" (quoted in Stamps 1995, 6). Spatialized time and temporalized space are the closely intertwined foundations on which capitalist society is able to regulate productive activities and hence to overcome various forms of resistance, both human and natural. From this segmentation of space and time, a materialist logic arises in which the most "rational" activities are those that accelerate processes, focus on short-range profits, and expand the scale of markets, production, and resource exploitation.

For Marx, *communication* is "a generic term to mask capital's lubricating operations" and refers primarily to information that accelerates production, increases the availability of goods and resources, mobilizes and captivates labor pools, and increases exchange value (La Haye 1979, 29). On this account, the cultural impacts of communication content (such as the words in a song or the plot of a book) are of little interest; what matters are the indirect effects produced by transformations in the mode and means of production through the reworking of space and time. In *Grundrisse: Foundations of the Critique of Political Economy*, for example, Marx argues:

> The more production comes to rest on exchange value, hence on exchange, the more important do the physical conditions of exchange—the means of communication and transport—become for the costs of circulation. Capital by its nature drives beyond every spatial barrier. Thus the creation of the physical conditions of exchange—of the means of communication and transport—the annihilation of space by time—becomes an extraordinary necessity for it. Only in so far as the direct product can be realised in distant markets in mass quantities in proportion to reductions in the transport costs, and only in so far as at the same time the means of communication and transport themselves can yield spheres of realisation for labour, driven by capital; only in so far as commercial traffic takes

place in massive volume—in which more than necessary labour is re-placed—only to that extent is the production of cheap means of communication and transport a condition for production based on capital, and promoted by it *for that reason*. (quoted in La Haye 1979, 125, emphasis in original)

This view of communication reflects a point in history before the telegraph, radio, television, and telephone had decisively wrenched communication apart from transportation (Carey 1988, 15). Marx conflated mobility and communication, but his contextual approach re-flects more than just the primitive state of nineteenth-century commu-nication technology. It also reflects a fundamental rejection of the idea that people act autonomously.

Like all actions, communication must, on Marx's account, be an out-growth of particular material conditions. "Not only do the objective conditions change in the act of reproduction, e.g. the village becomes a town, the wilderness a cleared field etc., but the producers change, too, in that they bring out new qualities in themselves, develop themselves in production, transform themselves, develop new powers and *ideas, new modes of intercourse,* new needs and *new language*" (from *Grundrisse*, quoted in La Haye 1979, 112, emphasis added). This argument from *Grundrisse* echoes a point made in the *The German Ideology*. "The fact that under favourable circumstances some individuals are able to rid themselves of their local narrow-mindedness is not at all because the individuals by their reflection imagine that they have got rid of, or intend to get rid of, this local narrow-mindedness, but because they, in their empiri-cal reality, and owing to empirical needs, have been able to bring about world intercourse" (quoted in La Haye 1979, 109–10). In Marx's view, communication is a reflection of a group's material situation, not a sphere for autonomous or creative activity. As Stamps indicates, he erases the line between communication and work: "For Marx, knowl-edge was something that emerged historically through the interac-tion of production processes and consciousness. Labouring was a kind of language, a practical language whose living vocabulary grounded all human thoughts and feelings" (1995, 18). By inflexible mechanical links, labor is language, space is time, and communication is production.

Like other Western philosophers, Marx grappled with the question of subject-object relations. Subjects (people) became objects when they lost control over the means of production. One particular subject was best situated, by virtue of its objectification, to comprehend *the process of objectification* in general: the proletariat held the answer to all the dilemmas of industrial capitalism because proletarians were the most thoroughly objectified subjects under capitalism. In Marx's view, the proletariat, as both object (of capitalism) *and* subject (thinking human), was predestined to discover the contradictions implicit in its own existence, to reject objectification, and ultimately to lead all of society in a revolutionary transformation.

In fighting to wrest control of the means of production, workers would necessarily gain consciousness of the ways in which their own powers had been alienated from them, and they would discover their role as "the revolutionary subject" and simultaneously find the language to articulate this role. Neither quiet reflection nor reasoned discussion nor formal education would drive the revolution; material conditions alone would drive it. For Marx, "the development of the proletariat's self-consciousness was an axiomatic element in the dialectic of history" (Alway 1995, 18).

Marx's argument is interesting because it clearly establishes a precedent for understanding meaning as discovered rather than created, almost like a mineral ore lying in the ground. It directly or indirectly inspired later, more nuanced interpretations of communication, including the writings of Claude Lévi-Strauss, his followers, and social theorists such as Antonio Gramsci, Michel Foucault, Louis Althusser, Pierre Bourdieu, Henri Lefebvre, and the Frankfurt School.[2] Many of these later writers built on elements of Marx's framework but were not as deterministic in their views, and communicators were once again included in the communication process.

2. For the least-recognized connection, see *Tristes tropiques* (Lévi-Strauss 1974, 49–50).

The essence of Marx's view of communication is that context determines content: insofar as people are positioned differently with regard to axes of social power, they must necessarily have different relationships to the field of meaning—different articulations and discourses. This idea, philosophically and politically radical, continues to resonate with modern scholars. Oddly, as social theory, feminism, sociology, anthropology, and human geography have adopted this idea, the relationship between social power and knowledge has been reversed: it is no longer the proletariat, but the bourgeoisie who are thought to have special access to knowledge. A cycle forms by which the rich get richer and the poor get poorer in terms of both material goods *and* the stock of available information. Marx's optimism is transmuted into pessimism; the inevitable utopia is replaced by an inescapable dystopia. In the words of Nigel Thrift,

> Thus, the social distribution of empirical (and practical) knowledge is associated with institutional nodes like home, school, university or office which form a set of points that selectively channel the life-paths of actors according to their membership of a particular social group. This channeling results in the acquisition of particular common kinds of knowledge (and the limits on that knowledge) that ultimately ensure the reproduction of that group as a socio-spatial entity. (1985, 388)

We see here a shift *within* the contextualist framework from optimism to pessimism, from romantic to tragic and ironic tropes guiding historical narratives. As noted earlier, this shift can be traced to interwar Germany and specifically to what has come to be known as the "Frankfurt School."

The Frankfurt School

Certain theorists in the first decades of the twentieth century rejected Marx's deterministic model of revolutionary consciousness, thereby turning consciousness into a problem rather than a given. The most important members of this circle were associated with the Institute for Social Research, a privately funded research institute founded in Frankfurt, Germany, in 1923 that later became known to English speakers as the "Frankfurt School." Physically located in the United States from 1933 to 1950, the members of the Frankfurt School saw American culture with eyes attuned to the manipulation of public opinion. Having seen the manipulation of opinion under Nazism, they were deeply concerned about public communication in American society.

Cross-pollinating Marx with Hegel, Kant, Heidegger, Husserl, and Freud, and following most directly the lead of Rosa Luxembourg, the Frankfurt School theorists asked what kinds of forces might either help or hinder revolutionary subjects in realizing their potential. The most notable of the Frankfurt School theorists were Max Horkheimer, Theodor Adorno, Walter Benjamin, Erich Fromm, and Herbert Marcuse. Although their theories were intended to breathe life into Marxism and adapt its theories to twentieth-century realities (including fascism, stable capitalism, and state socialism), they overturned Marx's romantic belief in the inevitable nature of social transformation. They depicted a system in which psychological and social conditions work together to "checkmate all the remaining hopes of social emancipation" (Piccone 1982, xvii). The exploitative nature of instrumental rationality was seen as so fundamental to Western civilization and its internalization so inevitable under capitalism that critical theory did "not even attempt to prefigure [a better] future by elaborating the mediations necessary to bring it about"; the intrinsic pessimism of this approach therefore closed off debate and eventually "swallowed social analysis" (Piccone 1982, xvi, xx).

Georg Lukács, an intellectual precursor to the Frankfurt School, was vital in problematizing the formation of class consciousness and linking it not only to material relations but also to ideologies (see, e.g., Lukács [1923] 1971). Lukács emphasized the proletariat's ability either to overlook or to exploit the opportunities provided by its objective position in society. His other vital influence was to foreground commodity fetishism as the "*specific* problem of our age" (quoted in Alway 1995, 18, emphasis in original). Following Lukács, the commodity, both in itself and as it was represented in advertising and popular texts, became an indication of the "reified" or overdetermined subjectivity. In Stamps's words, "the difference between systemic destruction [of forests] and the politics of [environmental] protest shades off into the difference between the plain coffee mug and the one depicting a stand of pines" (1995, 15).

The Frankfurt School theorists linked commodity fetishism to political manipulation and directed harsh critiques at mass media and popular culture, where commodity fetishism and political manipulation came together: "Modern communications media have an isolating effect; this is not a mere intellectual paradox. The lying words of the radio announcer become firmly imprinted on the brain and prevent men from speaking to each other; the advertising slogans for Pepsi-Cola sound out above the collapse of continents; the example of movie stars encourages young children to experiment with sex and later leads to broken marriages" (Horkheimer and Adorno [1944] 1972, 221). On this account, communications in the capitalist social context block communications that would otherwise bring people together to form families, communities, and nations. Not only popular culture but telecommunication and rapid transportation drive this erosion of political culture: "When visitors meet on Sundays or holidays in restaurants whose menus and rooms are identical at the different price levels, they find that they have become increasingly similar with their increasing isolation. Communication establishes uniformity among [people] by isolating them" (Horkheimer and Adorno [1944] 1972, 222). Here was a monumental irony—communication as a silence.

The two main contextual forces inverting communication in this way are capitalism and rationality. In Horkheimer and Adorno's argument, Plato's rationalistic forms of thought—abstraction, categorization, and quantification—derived their power from the unjust power relations of his society. Greek society, based on the labor of women and slaves, created thought forms that indirectly but inevitably served relations of exploitation, domination, and disenfranchisement. The particular link between Platonism and oppression is the treatment of individuals as members of groups, as specimens in a system of classifications. The specimen occupies a universe of things, in which the principle of universal substitutability leads to the use of specimens for purposes that lie outside the specimens' own interests. When classifications are elevated to the status of real objects, particular instances become unreal; the laboratory rabbit "as a mere example, is virtually ignored by the zeal of the laboratory" (Horkheimer and Adorno [1944] 1972, 10). Humans studied by science and managed by scientific bureaucracies are likewise reduced to the status of specimens. Instrumental rationality renders people's individual experiences meaningless, including their suffering: "When even the dictators of today appeal to reason, they mean that they possess the most tanks. They were rational enough to build them; others should be rational enough to yield to them" (Horkheimer [1941] 1982, 28). Those with power use rationality to justify their actions, demanding submission to a single measure of validity, which ultimately works like a religious doctrine to preempt and even vilify debates and alternative perspectives. Horkheimer therefore claimed in 1941 that "Reason, in destroying conceptual fetishes, ultimately destroyed itself" ([1941] 1982, 27). This silencing today takes the form of labeling "unacceptable" points of view as irrational, insane, rogue, treasonous, terrorist, communist, anti-Western, anti-American, or antimodern.

Horkheimer's colleague Adorno sought a solution in the fine arts, which do not pretend to be rational and therefore are "a promise beyond the suffering caused by the triumph of technical reason" (Bronner 1998). Using illusion covertly, rational discourses confirm the set patterns of "identity thinking" (a narrow worldview that supports social regimentation and domination). Art, in contrast, uses illusion overtly and thereby provides a means of transcending identity thinking. Of all the arts, music was Adorno's favorite. The syncopated beats of jazz held radical potential at one time, Adorno believed. But jazz had become too familiar, too formulaic, and now offered only the image of resistance: "the perennial fashion becomes the likeness of a planned congealed society" (Adorno 1981, 125). The overpowering interest in pleasing the audience encouraged a formulaic musical style that led, as he put it, to the "castration" of the audience (1981, 129–30), a comment that also shows Adorno's sexism. Adorno was wary of his own theories (depending as they do on rational argumentation), but he tried to bypass this problem through what he called a "negative dialectic," a form of communication designed to escape the trap of identity thinking.

Adorno applied this approach to the study of television among other texts. In "How to Look at Television," he recounted a few television plots that he considered typical in the way they defused class tensions and worked against revolutionary consciousness. These critiques are instructive because they reveal the limits of his particular contextual approach. In one show, the heroine, a schoolteacher, is unfairly penalized by the school principal, so she ends up going hungry. Nevertheless, she remains in good spirits and continues to make jokes. Adorno decoded the message of this comedy as follows: "If you are as humorous, good-natured, quick-witted, and charming as she is, do not worry about being paid a starvation wage. You can cope with your frustration in a humorous way; and your superior wit and cleverness put you not only above material privations, but also above the rest of mankind" (1981, 143). The program therefore discourages class struggle. In another program, some people are included in the will of an unknown benefactor, Mr. Casey. To their disappointment, it turns out Mr. Casey is a cat, and their inheritance is only cat toys. In the final scene, though, the toys are tossed in an incinerator, where they melt to reveal hundred-dollar bills tucked inside. From this tale, Adorno extracted the message, "Don't expect the impossible, don't daydream, but be realistic" (1981, 144), which also defuses class struggle. However, alternative interpretations of these two programs are possible: the teacher's wisecracks might be seen as challenging, denying, or subverting the hierarchy in which she must operate. The second plot might be seen as toying with capitalism's sacred tenets of self-interest, individualism, and abstract monetary value. Furthermore, this plot's reversals might be seen as prompting viewers to question the idea of money as the legitimate basis for a social hierarchy (because the money is destroyed and the "high-class" benefactor is actually an animal). But Adorno missed these interpretations and more generally overlooked the ambiguity of the televisual medium because of his preexisting expectations about context's determining power (Fiske 1987). These expectations make communication content seem fixed and determined by capitalism rather than subject to divergent interpretations founded in the multiple contexts inhabited by extensible viewers who live in and through the media they use.

We encounter in Frankfurt School theories a striking distrust of communication, a distrust that would later become a virtual hallmark of "social theory." The starkest expression of this view is the equation "Power and knowledge are synonymous," which was not, as many would guess, pronounced by Michel Foucault in the 1970s, but dates instead to Horkheimer and Adorno's brilliant treatise of the 1940s ([1944] 1972, 4). The reduction of knowledge to power grows directly out of the Frankfurt School's initial break with rationality, and several generations of social theorists have found ways to repackage the idea without moving much further. For example, Gillian Rose argued in the 1990s: "Disciplinary knowledge can define itself through its own ability to know only if there are others who are incapable of knowing" (1993, 9). Of the many echoes of Horkheimer and Adorno's "knowledge equals power" equation, the most deep and useful have come from Michel Foucault, whose central arguments I treat at greater length later.

In short, Horkheimer and Adorno believe the languages (verbal, musical, televisual, etc.) of an unjust and manipulative society must necessarily be thoroughly corrupted by their social context (which is viewed as monolithic and oppressive rather than plural and fragmented). Because any written critique of society must use a devalued or corrupted language, social intervention becomes a desperate mission: an effort to say something meaningful while skirting around language rather than embracing it, an attempt to express the inexpressible. In metaphorical terms, the critical theorist is like a doctor trapped in a dungeon, whose only tools for saving his fellow prisoners are torture devices that he must try to use as medical instruments. Beginning with Adorno's "negative dialectics," generations of critical theorists employed convoluted terminology and tortuous grammar in an attempt to rectify the supposedly contaminated logic of language. The possibility that language might be only partially politicized, that it might in some way correspond to a reality outside of or alongside of the social order (as in the correspondences previously suggested by symbols and signals) was not considered.

Hence, we arrive at the cynical formulation "false clarity is only another name for myth" (Horkheimer and Adorno [1944] 1972, xiv). As Paul Piccone argues, "The choice of a deliberately obscure mode of expression was not merely an unavoidable by-product of either the complexity of the phenomena [that the Frankfurt School theorists] chose to investigate or of their theoretical heritage, but was inextricably linked with the problematic character of their attempt to create a theory of emancipation in a context where all organized or organizable opposition had long since capitulated" (1982, xiv). Contextualist theory turned content into the enemy and thereby limited its own audience to those with a special (i.e., exclusive) education.

Instead of starting with messages and working toward interpretations as we do in analysis of content, the Frankfurt theorists began at the level of society, then deduced the nature of *possible* communication acts from this view of society. Ironically, by doing so, they reduced the participants in communication to caricatures and thereby adopted the Platonism they deplored because these caricatures were ideal types rather than actual instances of humanity. They also adopted an attitude toward language that undermined their own ability to communicate: "internal complexity and obscurity [of writing] were to be in themselves guarantees that the process of decoding explosive theoretical contents would shatter the reification resulting from automatic readings and conceptual commodification" (Piccone 1982, xiv). This translation of contextual critique into opaque academic writing became almost a ritual.

Two members of the Frankfurt School took a more optimistic view of communication. Walter Benjamin's "The Work of Art in the Age of Mechanical Reproduction" celebrated the intervention of mass production in the artistic process, arguing that new communication technologies were democratizing artistic production and enabling "the original to meet the beholder halfway, be it in the form of a photograph or a phonograph record" ([1969] 1986, 30). Herbert Marcuse shared Benjamin's optimism about communication, but for reasons more closely linked to Adorno's interpretation of culture.

For Benjamin, as spatial barriers to communication fell, so too did social barriers: "The cathedral leaves its locale to be received in the studio of a lover of art; the choral production, performed in an auditorium or in the open air, resounds in the drawing room" ([1969] 1986, 30). This detachment from place, which Giddens would later label "disembedding," had radical political potential, Benjamin believed. Art originally served the purpose of social repression through its ritual associations between power and "the good," but art itself was undergoing transformation as a consequence of reproductive technologies such as photography and sound recording. The inexpressible quality of uniqueness traditionally associated with the "work of art," which he called its "aura," is grounded in unequal power relations in the material world and in space, enhancing their ability to perpetuate oppression. Mass-produced technologies such as the hand-held camera and phonograph devalue the aura and put the power of art back into the hands of ordinary people: "the distinction between author and public is about to lose its basic character" because at any moment "the reader is ready to turn into a writer" (Benjamin [1969] 1986, 39). A third facet of Benjamin's optimistic vision was that in a paradoxical fashion the technologies born of rationality were creating a world of juxtapositions that challenged rationality and rationality's social control, particularly in the phantasmagoric images of mass consumption and the modern marketplace.

Marcuse on his part critiqued his Frankfurt School mentors, arguing that "Even in its most distinguished representatives Marxist aesthetics has shared in the devaluation of subjectivity" (1978, 6). His approach emphasized the receiver's subjective experience of art and popular culture: "The transcendence of [the receiver's] immediate reality [by encountering the text] shatters the reified objectivity of established social relations and opens a new dimension of experience: rebirth of the rebellious subjectivity" (7). In terms borrowed from Freud, he argued that the pleasure of art, even "decadent" art, was produced by "desublimation"—that is, by getting back in touch with desires that had been systematically misdirected by society for the purposes of economic exploitation. Hence, popular culture could indeed be progressive if its psychological effects encouraged personal empowerment. Here he developed one of Adorno's ideas. The subject's delight in communication arises from the glimpse it provides of a world containing greater free-dom and happiness than the present one. And because freedom and hap-piness are "the ultimate goal of all revolutions" (69), the experience of these emotions even through fiction or fantasy can help subjects deob-jectify themselves and orient their energy toward making social changes.

▲ ▼ ▲

Marx provided the fundamental insight that communication often serves instrumental goals by creating spatialized time and temporalized space. His concept of revolutionary consciousness was inadequately theorized, and the Frankfurt School began to address this oversight by showing how communication maintains the stability of an oppressive society. This literature raised the possibility that some aspects of com-munication are predetermined by languages—contexts that set strin-gent constraints on the type of consciousness people may have. It suggested, however, that forms of communication lying outside ra-tional argumentation (such as music and other systems of symbols and signals) may have the potential to motivate social struggle, despite lan-guage, through the new forms of consciousness they may engender. Without the separation of sign, signal, and symbol, it is not clear how common, how powerful, or how rapidly exhausted such innovations may be. Nor is it clear just how writings in social theory can assist with the formation of a radical, resistant, or revolutionary consciousness. Al-though content-based theories help respond to the first issue, Jürgen Habermas's theory of communicative action responds to the second. It provides some sense of *(a)* how individual communicators work to af-firm or challenge existing worldviews from within the preexisting con-straints of language, *(b)* how progressive writing (such as social theory) is itself possible, and *(c)* what kind of physical and technological con-texts facilitate socially constructive communication.

Jürgen Habermas and "Communicative Action"

Habermas draws on the theoretical foundations of the Marxian philosophy of consciousness, as developed by the Frankfurt School, and combines these theories with foundational writings of symbolic interactionism, in particular the ideas of George Herbert Mead as well as the sociological theories of Talcott Parsons and Émile Durkheim and the later philosophy of Ludwig Wittgenstein. From this epistemological stew, he has produced a theory not of consciousness or communication, but rather of a particular kind of action he calls "communicative action." In theorizing about action rather than consciousness, he aligns himself with sociology, yet many of his key concepts, such as the "colonization of the lifeworld" are strongly supportive of Frankfurt School philosophicosocial criticism. The shift from a theory of consciousness to a theory of action separates Habermas from the rest of the Frankfurt School and makes his approach inaccessible (and therefore unacceptable!) to the vast majority of current scholars in "critical theory" whose work continues to bear an acknowledged or unacknowledged debt to the rest of the Frankfurt School and primarily to the six-decades-old writings of Horkheimer and Adorno.

Communicative action is not language alone or the sum of various texts and discourses, but rather a process including speech and language as well as actions that have real-world effects. The theory of communicative action situates the exchange of signs and symbols as one facet of an overall human-environment and individual-society relationship. What is indicated is a very broad process of synthesis: "Communicative action relies on a cooperative process of interpretation in which participants relate simultaneously to something in the objective, the social, and the subjective worlds, even when they *thematically stress only one* of the three components in their utterances" (Habermas 1987, 120, emphasis in original). This argument rests on the deduction that a communicative act is *understood* only when participants are able to fit its meaning into ready-made associations between subjective, objective, and social phenomena in a particular situation. But the "horizons" or epistemological scope of any communication situation shifts with every successive articulation, and "no participant has a monopoly on correct interpretation" (Habermas 1984, 100). People struggle to define the boundaries of reality, but the consequence of a breakdown in interpretations is not a step "into a void" of unreality, but rather a shift among various alternative understandings corresponding to varying "horizons" in virtual and physical spaces (Habermas 1987, 125). In other words, communication depends on context, but constantly *redefines that context* in what is never a purely social or political way.

Communicative action is much more than communication, as communication is generally conceived. Communicative action is the achievement of the concrete goals of *(a)* social coordination, *(b)* personal expression and individuation, and *(c)* accommodation both to and of the material environment. This synthetic view of communication implies that communication should not be analyzed psychologically, philosophically, or sociologically, but rather from a combination of these perspectives because it operates at the juncture of the subjective world, the objective world, and the social world. The theory accordingly binds together not only ideologies, beliefs, knowledge, and norms, but also natural phenomena and various limitations imposed on human action; it is both descriptive and normative. The normative dimension arises insofar as communicative action entails action oriented to reaching self-understanding and mutual understanding. This speech situation is an ideal, however, so the fundamental question of Habermasian "discourse ethics" is to what degree actually existing speech situations approximate this ideal. This question leads directly to traditional geographical notions of place as well as to new debates on virtual place.

Like Horkheimer and Adorno, Habermas frames history as a process of rationalization, but in his view rationalization contains some positive developments, including *progress* in the strong sense of the word. For this reason, his "critical theory" differs markedly from the writing of many other authors in the critical theory tradition. Opportunity for individual autonomy is both a precondition and a product of what Warren calls "democratically empowered discourse"—in other words, communication that is rational and takes place in a favorable context (Warren 1995, 178). Rationality increases, according to Habermas, insofar as communication is framed in contestable validity claims corresponding to the three "worlds." First, the sender's subjective sincerity is judged by receivers: Does she honestly mean what she says? Is she signaling her intent in a way that is sincere and free of self-deception? This type of validity claim presupposes that the senders and receivers of a communication share some elements of subjective experience (a supposition based, naturally, on a measure of trust). Second, the objective truth of the sender's statement is assessed by receivers: Is she right? Does her meaning conform to the real world as I and others understand it to be? Can her claim be true? Here we depend on the motivated quality of symbols, especially symbols built up from signs (words) carefully organized to reflect some aspect of experience. Third, the intersubjective moral or ethical implications of the sender's meaning (symbolic content) are judged by receivers: What relationship is the speaker assuming relative to others by addressing her communication in this particular way? What social norms, values, and expectations is she following, reinforcing, or violating by saying it in this way?

Rational statements are open to arguments based on any or all of the previous validity claims. The history of Western society is, on this account, a progressive increase in rationality because unquestionable communications—ritual, myth, and magic—have been replaced by debatable communications. This means that validity claims can increasingly be tested, verified, clarified, and refined. We must of course trust the sender of a communication before we are willing to consider seriously his or her validity claims and offer a serious reply. We must be *willing* to craft a reply that in some sense does more than echo other articulations we have heard—which demands a degree of creativity in our use of signs and symbols. We must be willing to extend a certain kind of trust to all other parties in communication that is based, minimally, on the willingness to give the speaker the benefit of the doubt that she meant more than she could say or said less than she meant, so that if our disagreement appears total, it is perhaps only partial, and we can find some common ground. Qualities of trust and creativity are necessary for rational communication to "work."

In addition to rationally debatable validity claims, Habermas's *ideal speech situation* stipulates that every person with the competence to communicate is allowed to participate, to question any assertion made by others, to introduce any assertion, and to express his or her attitudes, desires, or needs. Constraints to such participation must not exist either within the communicating subject or external to him or her. Again, we see that Habermas's communicative action is founded on the assumption of trusting, creative subjects who communicate in symbols and signals built up of signs. On this account, it is not impossible to transcend biases and conflicts of interest in a temporary and limited way. There is a social framework for rational noninstrumental communication. Content need not be tortured in order to escape contextually predicated social oppression.

Let us consider what this means in an example of ordinary communication, one of the few provided by Habermas. In Habermas's example, the foreman on a work crew tells one of the junior crew members to go pick up some beer and to hurry so as to get back in a few minutes (Habermas 1987, 121). The junior crew member (if he is from the foreman's culture) will probably understand *(a)* that he is being told to do something by a superior in the work setting and thus has few alternatives other than compliance or quitting the job, *(b)* that the beer is for the upcoming midmorning break, *(c)* that there is a shop within a few minutes' walk that sells beer, and *(d)* that the foreman is sincere and is not making the request as some kind of practical joke, trick, or metaphorical reference to something other than beer. "Understanding" the request in this case amounts to accepting an undetermined number of *different validity claims,* including an *intersubjective* claim relating to the social relationship between the speaker and the listener, an *objective* claim relating to the spatiotemporal relationships of work site and shop, an *objective* claim relating to the shop's inventory, and a *subjective* claim relating to the desires and intentions of the person making the statement.

Communication cannot be understood simply as a structured mental construct (subjective fact) or as a set of social conventions (intersubjective fact) or as a model of nature (objective fact). It is a fact in all three senses: a means of bringing together the mental and the social worlds for the purpose of coordinating collective actions in the social and the physical worlds and for the purpose of defining the self: "With every common situation definition [people] are determining the boundary between external nature, society, and internal nature; at the same time, they are renewing the demarcation between themselves as interpreters, on the one side, and the external world and their own internal worlds, on the other" (Habermas 1987, 122). Linguistic products are therefore not arbitrary, like the signs that compose them, but are motivated wholes, symbols of several aspects of reality. To see this correspondence, however, we must simultaneously consider social, psychological, and physical reality. A single text, with its sequence of signs, corresponds in some way to all three realms, and if these realms are only considered separately, its validity, *which is to say its reality,* cannot be evaluated.

The symbolic representations circulated by communicative action in places and virtual places constantly define and redefine what is within the "horizon" of the participants' sense of place—a range of sensation and potential action. "Here" constantly shifts through extensible subjects' communicative action. We can extend Habermas's explanation as follows. If the worker mentioned in the example is unaware of the social hierarchy (intersubjective condition) underlying the foreman's authority, he may say, "OK, I'll get the beer, but it's your turn tomorrow," after which the foreman's annoyance, outrage, or laughter may prompt a better intersubjective understanding about the worker's lack of authority in this situation. If the worker is unaware of the proximity of a shop selling beer (objective condition), he may ask, "Where do I go?" prompting verbal or visual directions to the shop. If he is uncertain about the foreman's intentions (subjective condition), he may ask, "Are you serious?" prompting a reply ranging from "Quit complaining and hurry up!" to "Of course not, we don't drink on the job, mortarhead!" or gestures ranging from a nod of the head to a dismissive wave of the hand (communicating in the realm of symbols). If trust is misplaced or overextended by trickery, creative responses are sure to follow.

Communicative action, then, although informal and guided by unstated rules, works to produce the world in its ontological fullness. Each communicative exchange, however mundane, draws in elements of reality from beyond the "horizon" of the situation: matters of fact, social norms, and subjective states. This is what Habermas means by the observation that "Situations do not get 'defined' in the sense of being sharply delimited. They always have a horizon that shifts with the theme" (1987, 123). We might also say that the subject's extensibility fluctuates from sentence to sentence. Furthermore, the validity of any given articulation rests on whether one's audience can identify what aspects of the objective, subjective, or intersubjective world are presently being symbolized. Misunderstandings, too, can involve disjunctures in the definitions of any of these realms.

Like other Frankfurt School theorists, Habermas approaches the study of society as a political project, only his project is not to develop a philosophy of alienated consciousness but to restore vitality to what he calls the *lifeworld*. The lifeworld is the sum of social coordination constructed through personal communication, argument, and consensus building. Under Habermas's broad interpretation, it includes virtually all person-to-person communications in words and images as well as the norms, expectations, and understandings that people employ in an ad hoc and spontaneous manner to coordinate their actions with others. The *lifeworld* is threatened by the *system,* a realm of social coordination based on abstract, rule-driven modes of interaction, such as money and law, and deployed within the organizational frameworks of capitalism and the state. In Habermas's terms, the lifeworld consists of "mechanisms of coordinating action that harmonize the *action orientations* of participants," whereas the system consists of "mechanisms that stabilize nonintended interconnections of actions by way of functionally intermeshing *action consequences*" (1987, 117, emphasis in original). The lifeworld meshes signs, symbols, and signals, whereas the system depends mainly on signs. The system's threat to the lifeworld takes the form of "colonization"—replacement of interpersonal communications oriented toward mutual understanding by impersonal, abstract communications oriented toward strategic, instrumental control.

Human agency is central to Habermas's theory insofar as the differing means of coordinating action lead to radically different types of social coordination with different moral implications.

Habermas breaks from the Frankfurt School on the topic of rational thought. Rather than offer a blanket rejection of rational thought, he insists that rationality has led to progress through the elimination of magic, the reduction of prejudice and superstition, and the increase of material prosperity and security. He sees rationality as social coordination that facilitates the achievement of personal and collective objectives, both by aligning goals with values (subjectively rationalizing action) and by improving the efficacy of methods (objectively rationalizing action). The coordination of goals, values, and methods in collective action depends on rational communication, which is to say communication that employs criticizable validity claims relating to subjective, intersubjective, and objective realities. He argues that we "call a person rational who interprets the nature of his desires and feelings [*Bedürfnisnature*] in the light of culturally established standards of value, but especially if he can adopt a reflective attitude to the very value standards through which desires and feelings are interpreted" (Habermas 1984, 20). Habermas breaks with the Frankfurt School and virtually all of contemporary critical theory in identifying the problem of modernity not as rationality per se, but as an overly narrow definition of rationality that excludes emancipatory (intersubjective and altruistic) interests and focuses only on technical and practical (objective) interests.

Horkheimer and Adorno would argue that people's trust in what Habermas calls "culturally established standards of value" fatally compromises communication. Habermas has somewhat more hope for what is culturally established, at least when it is constructed of criticizable validity claims. He is not without sympathy for Horkheimer and Adorno's concern that communication that appears rational may nonetheless support extreme social injustice, but he narrows the field of critique from rationality writ large to a particular kind of destructive rationality—the impersonal, practical thought manifested in abstract systems such as money and law. Modernization has, in his view, brought about the progressive "mediatization" of social relationships: the penetration of the lifeworld through impersonal "steering media" such as money and law. These (sign-based) steering media coordinate activities (legally or monetarily) without the need for personal assertions of value by participants.

Mediatization produces many benefits when properly constrained—that is, prevented from creating unwarranted intrusions on the lifeworld. Most obvious of these benefits are material control over the environment and more effective social administration; "every modern society, whatever its class structure, has to exhibit a high degree of structural differentiation," and steering media make such specialization possible (Habermas 1987, 340). Only when *instrumental rationality* begins to "colonize" the lifeworld do problems arise. Impersonal coordination replaces autonomous action, and people are both internally and externally constrained. Habermas clearly assumes that colonization of the lifeworld is merely an unfortunate by-product (not an essential flaw) of progress, and he therefore attempts to separate colonization of the lifeworld from positive consequences of progress and rationalization in particular. This approach allows us to develop an overall program for improving a communication system rather than simply critiquing communication. It also helps distinguish between various communication situations such as newspaper advertising, political propaganda, fine art, television comedy programs, and protest marches. The goal of any progressive communication must be to treat normative concerns in a deliberative fashion, through *rational discussion*. The realms of individual socialization where norms should be contested, such as the school and the home, and the "public sphere" of political debate, must not be dominated by steering media and thereby deprived of the open-ended communication that provides for moral thought and behavior.

The objective in describing an ideal is to understand one pole of an actually existing continuum. By understanding this pole, we may discover that things that had previously escaped our attention become evident. Although Habermas shares with other proponents of critical theory the goal of emancipation, he is more creative in thinking through the conditions of emancipation rather than simply the conditions preventing emancipation. In this context, we should recall Habermas's interpretation of social science as the branch of scholarship dedicated to emancipation, guided not by the technical and practical objectives suited to natural science and subjects such as anthropology and history, but by an attempt to reveal distortions of subjectivity and to suggest means for overcoming them (Dryzek 1995, 98). The problematic rationality identified by the Frankfurt School appears now as a specific kind of rationality, a rationality characterized by the fact that it silences communication about subjective experience and closes off debates about intersubjective questions of justice. The subjective and intersubjective are also, not coincidentally, most often expressed through symbols and signals laminated onto verbal sign communication.

This argument has the profound implication of resituating the debate over morality and ethics from questions of justice and rights to questions of procedure and social context. It is a deontological rather than a teleological approach to morality (White 1995, 10). With regard to the existing social science literature, Habermas broadens the realm of moral concern from questions of economic *structure* to questions of political and legal *process*. In regard to communication, he is advocating a rational political process, but because he is not proposing a *standard* of rational thought or any particular method of judging what ideas are more or less rational, he escapes the pitfall of earlier proponents of rationality. Rationality is not something for an authority to determine and uphold as a standard, but rather a characteristic of dialogue itself—specifically the communicators' freedom from threats, coercion, and relationships of dependency (economic or otherwise). A rational basis for public life depends on organizing social relations "according to the principle that the validity of every norm of political consequence be made dependent on a consensus arrived at in communication free from domination" (Habermas 1971, 284). Therefore, the "answer" determined rationally (in the broad sense) is always necessarily provisional and temporary, but does nonetheless bear a nonarbitrary tie to reality as do other, simpler symbols.

But to what extent do such communication situations actually exist? Habermasian theory supports an interest in place that has long motivated human geographers. In fact, when defining the nature of the ideal speech situation, Habermas rejects communications that do not occur in places, face to face. Kellner demonstrates that this requirement is unnecessary and that it makes sense to conceive of the public sphere as "a site of information, discussion, contestation, political struggle, and organization that includes the broadcasting media and new cyberspaces as well as the face-to-face interactions of everyday life" (2004, see also Kellner 1995). Democratically empowering places exist only insofar as the *potential* for domination is held at bay in particular places and virtual places (distanciated contexts for communicative action). A place or virtual place must permit people to subordinate power relations temporarily to other organizing principles such as trust. The power of certain physical places or virtual places to exclude hierarchies of social power temporarily and conditionally is a crucial assumption if we are to adopt a Habermasian approach. Rather than seeing one context—capitalist society—that generates a devalued language, we must see many bounded situations in which languages of various kinds (verbal, musical, pictorial, televisual, etc.) may be entrained by various social actors to symbolize a potentially infinite range of debatable realities. These realities must be up for "debate," but in a sense that is broader than just sign-based, objective debate. Subjectivity must be discussed if a discourse is to reflect reality.

We can go further and assert that what makes colonized elements of the lifeworld "dysfunctional" is that they no longer provide the bounded contexts for identity formation that are required if individuals are to develop into fully moral agents (Warren 1995). Without places—physical and virtual—in which to carve out a temporary approximation of the ideal speech situation, the subject has little choice but to adopt an amoral identity that is content to avoid moral decisions and simply to "follow the rules." This construction of self cannot have much self-understanding or understanding of others. It is deaf and blind to their self-expressive subjective signals of economic frustration, for example. The exercise of trust, altruism, creativity, and perhaps even love is curtailed.

Morality therefore emerges out of a particular relationship between people and their communication contents and contexts, rather than from universal or a priori moral precepts.

This concern with the conditions necessary for the constitution of autonomous agents in and through communicative action dovetails neatly with the theory of the extensible self. Celebrations of new technologies generally assume people to be empowered inasmuch as they incorporate new communication technologies into their lives and acquire the how-to knowledge that lets them find their preferred experiences in and through these media (Bianculli 1992; Lyon 1988; Toffler and Toffler 1995). What this naïve technological boosterism overlooks is the quality of social interactions people can have in and through particular contexts as well as the need for social intervention in the form of teaching people how to appropriate virtual places and to use them to "articulate their own experiences and interests, and to promote democratic debate and diversity" (Kellner 2004). These interactions inevitably reverberate in real-world contexts, shaping face-to-face interactions in complicated ways (Meyrowitz 1985). Communication therefore is a fundamentally geographical process involving every aspect of context: space and place (or geography), the communication media at one's disposal, and the enveloping social structure and institutions. The interactions between physical and virtual contexts are so complex that it is counterproductive simply to proclaim the corruption of meaning by either a social or a technological context of communication alone.

Another useful guidepost to analysis is provided by Michel Foucault, although in practice the Foucauldian method has been less nuanced that its original conception. Foucault's most important insight was the substitutability of institutional for physical contexts of communication. Thus, the "panopticon" can be either a physical structure, a series of cells encircling a central observation post, or an institutional framework subjecting inmates, patients, workers, or pupils to constant surveillance. In the latter situation, a person "becomes the principle of his own subjection" (Foucault 1979, 203). The commonality resides in the way physical environments and institutional frameworks organize perception of one group of people by another, perception of one individual by another, and perception of a person by himself or herself. In this organization lies the "humble" principle by which power is exerted over individual lives. Yet there is something stifling about Foucault's underlying premise regarding power. If power can be reduced to "a certain concerted distribution of bodies, surfaces, lights, gazes; in an arrangement whose internal mechanisms produce the relation in which individuals are caught up" (Foucault 1979, 202), then by what means can perception of any sort be disentangled from power relations so that we understand perception as an expression of autonomous action on the part of the individual?[3] How can social analysis accommodate contextuality (as a general principle) while simultaneously doing justice to the subjective sense of autonomy shared by the reader and the author? And how can scholars avoid a kind of elitism in which they explain away the possibility that "people" (meaning *other* people) can act and communicate freely? Refusal to recognize oneself is a kind of narcissism and gives rise to both irrationality and oppression.

This problem arises in connection with another author scarcely less influential than Foucault in the social sciences. Edward Said (1978) has revealed a fundamental connection between knowledge and power in geopolitical relations, coining the term *Orientalism* to describe the combination of beliefs, generalizations, and implied policies imposed by Westerners on a vast region of the non-Western world. Perception and actions were inextricable as *Oriental* came to imply passivity and submissiveness, which was exactly what the West hoped for, as well as inferiority, which was the West's moral justification for conquest and control. Said avoids the issue of anti-Western definitions of the West, from China's arrogant dismissal of Europeans as barbarians to the Islamic world's condemnation of the West as lawless and chaotic through the concept of *Dar al-Harb*. Perhaps the Othering of the West by the East was less apparent because of the absence of international forums similar to Western media in which such a discourse could have circulated. Factors such as the growth of literacy, increased travel, and growing access to communication technologies have reduced this disparity, creating greater opportunities for prejudiced Eastern worldviews to be deployed in retaliation against the prejudiced Western worldviews. Perhaps these virtual and physical spaces might also be used to build dialogues between East and West, supporting cultural and political rapprochement or at least a modicum of mutual understanding. Simply condemning Orientalism does not take us in that direction.

In the next section, I explore the ideas introduced thus far in the

---

3. For a discussion of how new communication technologies operate as instruments of surveillance that thereby rework personal identity, see Curry 1997.

chapter by taking us on a tour, as in the previous chapter. This time our journey will be in virtual rather than physical space. First, though, I should recapitulate the points made thus far in the chapter.

▲ ▼ ▲

A materialist interpretation of communication attempts to trace the links between contexts of communication and the meaning of the communication act. Context is a flexible concept that includes physical structures, social institutions, and technological apparatuses. Despite their manifest differences, all three types of context work in similar ways to structure perception between groups and individuals, defining knowledge and power. One type of context, capitalism, has overwhelmingly dominated contextual approaches thus far, as indicated by the predominance of Marxist and Marxian perspectives in this discursive field. Whereas Marx gave this type of context determining power, the theorists of the Frankfurt School saw communications as somewhat autonomous because of their ability to shape subjective awareness, sense of self, and knowledge. These theorists' conclusions were often bleak: capitalism generated debased languages (including not only verbal languages but music and art as well), and the commercial imperative led to a general tendency for all capitalist communications to objectify the world. Yet a kind of grail beckoned in the form of new technologies and new languages (broadly defined), as suggested by Benjamin in connection with the camera and by Marcuse and Adorno with respect to music. New languages and media might allow the communication of ideas that were previously silenced.

More recently, similar arguments have been advanced in connection with electronic media. John Fiske (1987) argues that the peculiar visual "language" of television provides an open text supportive of conflicting interpretations and heterogeneous worldviews. Joshua Meyrowitz (1985) argues that the language of television (specifically its relatively accessible character) and its ability to cross barriers between public and private spaces destabilized traditional power relations, reconfiguring the meaning of gender, age, and political power during the second half of the twentieth century, and Melvin Webber (1964) sketches the geographical implications of this process. This sense of an "opening" for communications employing new languages and media suggests that the Internet might similarly provide at least a temporary opportunity to voice dissent and to upset taken-for-granted worldviews such as Orientalism and corresponding anti-Westernisms. Thus, Mark Poster claims that "The age of the public sphere as face-to-face talk is clearly over: the question of democracy must henceforth take into account new forms of electronically mediated discourse" (2001, 181). If this discourse remains fragmented into "virtual ethnicities" and "cyber tribes," then perhaps it is not such a limitation because the competing ideal of "the public" has always in fact served to disguise social fragmentation and exclusion; the unveiling and reworking of this eternal fragmentation online is potentially a step toward greater mutual recognition.

Surfing a Small World

Our journey starts with a Web site I created for my introductory human geography course. The goal will be to make a "round trip" to India and back following links from this site. Our trip ranges widely in space and in virtual spaces of symbolism, but it is bounded in time. The sites we visit and the links between them were as I describe them in summer 2003. A year later they had changed significantly. Figure 4.1 maps our trip through virtual space. Pages generally accessed only through links from a "parent" site are encapsulated within the box of the parent site. As suggested by the font size, the University of Texas at Austin Web site, the Global Hindu Electronic Networks (GHEN), and Yahoo.com are the most important entities in this diagram. The level of connectivity of these organizations within the Web, measured on AltaVista.com, is indicated in crude fashion by the font size.[4] Figure 4.2 shows a more precise comparison of inward-directed link counts from 2003 and 2004. We can see that the number of links *to* the www

4. The number of links to a chosen site—for example, www.yahoo.com—were calculated by entering the "link" command in the AltaVista search engine—for example, link:www.yahoo.com, which yields an approximate count of all sites with links to www.yahoo.com.

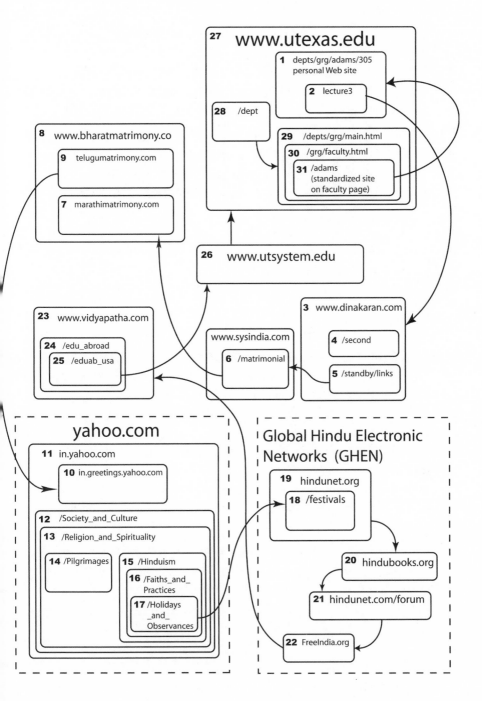

4.1. Round-trip Web path taken in summer 2003, from the author's class Web page.

Connectivity of Selected Sites

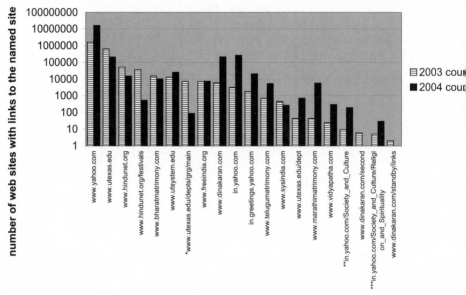

4.2. A measure of nodal centrality on the World Wide Web, measured in terms of the number of Web pages with links to the selected URLs. Numbers were calculated during summer 2003 and fall 2004 using the AltaVista search engine.

* The site www.utexas.edu/depts/grg/main was replaced in 2004 by www.utexas.edu/cola/depts/geography.

** The site in.yahoo.com/Society_and_Culture was replaced in 2004 by in.dir.yahoo.com/Society_and_Culture.

*** The site in.yahoo.com/Society_and_Culture/Religion_and_Spirituality was replaced in 2004 by in.dir.yahoo.com/Society_and_Culture/Religion_and_Spirituality.

.utexas.edu site or to any of its affiliated departmental or program Web sites and of links to the various Yahoo or GHEN sites is many factors of ten greater than number of links to a relatively obscure site such as the online class syllabus for my college course. In the terminology of network theory, the most popular sites (those linked to by many other sites) are *hub* sites, sites that provide access to many other sites through categorized links. The least popular sites are *leaf* sites that form dead ends or virtual dead ends in the topological space of the World Wide Web.

1.

The page from which we depart on our journey is a fairly typical on-line college course syllabus (one of my own). The black background with yellow and white lettering is a stock color combination in the Web design software. Aside from an image of the globe, all this page contains is information about the class and the professor, along with links to twenty-one lectures, two projects, and three review sheets. Although the basic page is similar in content to a traditional paper syllabus, it is differ-ent because the links transform the syllabus into the hub of a tiny net-work of pages so that it is not just an object standing alone. As usual, context issues involve power. Through its links and its online presence, the webbification of the syllabus reworks power relations between and within academic actors (students and instructors).

Whereas a paper syllabus would be handed out on the first day of class and dutifully guarded by students throughout the semester, an electronic syllabus is always available online. Whenever students want to check a due date, they can call up the syllabus on the same screen they use to download music, read e-mail, write papers, and play games. That the Internet can be accessed from any place is by now a banal observa-tion, but what that means with regard to how the student "becomes the principle of his own subjection" (Foucault 1979, 203) is interesting. When the syllabus contains hypertext links to project descriptions, re-view questions for exams, lectures, and other supplementary materials, the problem (and the excuse) of forgetting to bring the right paperwork home for the weekend is eliminated. If a student or a student's parents do not have Web access, there are online computers at the library, at friends' houses, and so on. A large part of the responsibility college stu-dents used to have is now cast off like a reptile skin, yet new types of re-sponsibility can emerge to take their place, such as the responsibility to have a computer account with Web access included and the obligation to check e-mail regularly.

This is not necessarily a boon. The minor liberation from responsi-bility may send a signal to students that results can be instantaneous and that technology can substitute for discipline, ultimately making it more difficult for them to acknowledge the responsibilities that remain. If the term paper handout can be obtained instantaneously from the Web, then perhaps so can the term paper itself. When students try this short-cut, the Web also makes it easy for professors to run a quick check on suspicious papers. The unequal power relations between student and in-structor are not eliminated by the new technology, but they do take on new manifestations.

Digging deeper into the context of this communication, we find further ramifications for the power relations in the class. It would theo-retically be possible to study the online lecture notes instead of attend-ing the lectures. We can see that in response to such a scenario, the professor (me) has designated 25 to 30 percent of the total credit in the class as "participation credit." This practice affords an opportunity to reward attendance, classroom discussion, or participation in in-class projects. This somewhat Machiavellian grading policy prevents technol-ogy from serving as a stand-in for the professor. In essence, a technique (grading) compensates for the disruptive influence of a new technology. The combination of new technology and new technique conserves the power of the instructor over the space-time of instruction, but does not simply repackage the same old power relations. Students find they must contribute to in-class discussions in order to pass the class. Although the classroom is not an "ideal speech situation" in the Habermasian sense, more practice speaking resolves some of students' anxieties about public speaking, in turn constituting them as knowledge creators rather than as knowledge absorbers. In this case, the adaptation to a Web technology has inspired a second adaptation, toward greater stu-dent communication skills in place. If the first wave of adaptations to new communication technologies can occasionally be predicted, it is much more difficult to predict such second-wave adaptations. Whether we see oppression or liberation in the new communication situation, we must be clear that the technology does not achieve this transformation on its own, as an actor in its own right. Technology cannot act au-tonomously and does not have a will, free or otherwise. The sociospatial transformation occurs as a result of a cascade of adaptations and appropriations made by people adapting their agency to a constantly changing technologically mediated environment.

2.

We leave the class syllabus page by clicking on the first link in a list of lectures, which happens to be Lecture 3. (Lectures 1 and 2 lack formal notes.) Lecture 3 is called "Elements of the Indo-American Bridge-space." From here we will embark on a short trip to India and back. Owing to the "small-world" character of the Web, we are able to go virtually anywhere on the Web in less than twenty steps, although research indicates that we will never be able to access some parts of the Web via links (Albert, Jeong, and Barabási 1999; Barabási 2002, 165–69).

The Web page for Lecture 3 turns out to be nothing more than a list of links with a few attached images. Like a stack of quotes on sheets of paper, these links appear to be nodal ideas around which the lecture is constructed. Not much can be learned from this page (because I am bracketing out my own knowledge of the site) but from a PowerPoint lecture linked to the page, we learn that *(a)* a particular community or set of related communities can create a virtual space to serve its needs; *(b)* such a virtual space can function as a bridge between different times and places, preserving cultural interactions despite physical separation; *(c)* such a "bridgespace" can have a political role relative to constructions of nation and ethnicity; *(d)* such a bridgespace can function economically as a marketplace for goods and services and as a channel for investment; and *(e)* a bridgespace can simultaneously slow and accelerate cultural diffusion processes. The implication is that the links provided on the class lecture Web site (to sites by and for immigrants from India to the United States) are illustrations of a virtual social space that intervenes in cultural processes among a population split between a homeland (India) and various destinations temporarily or permanently adopted by persons of Indian descent.

3.

Picking a link at random, we leave the lecture page en route to www.dinakaran.com. This page's main "content" is a choice of continuing in English or Tamil, but before making the choice we look over this simple site and follow the banner advertisements to see what they are about. The heading on the site reads "Keep in touch with developments in Tamil Nadu," suggesting that the site's main audience is intended to be members of the Tamil ethnicity living in and outside of the southeastern Indian state of Tamil Nadu. At the bottom is a robin's egg blue bar with the words "Hot Politics, Hot Cinema, with your Hot Coffee—Daily," suggesting that the Internet can be inserted into a daily routine traditionally associated with the morning paper. Prominent on this page is the recommendation "To view this site's content, Kindly download the Amudham font and install it in your machine."

A technological barrier that might exclude less technologically adept Tamils from this site (lack of the proper font) is minimized. But the site remains exclusive in another way simply by the use of Tamil (albeit mixed with English). Visitors who do not read Tamil (a group that includes non-Tamils and some well-educated Tamils because English-language schools are common among this group) will find parts of the site difficult or impossible to understand. The technological context includes and excludes, yet by supplementing other forms of inclusion (e.g., newspaper and face-to-face communication), it affords more people more opportunities to communicate.

Another aspect of context, capitalism, is evident on the entry page. Here we find five banner advertisements: two for America Online, one for a telephone calling card company called VDT, one for a bank called ICICI, and one with job applications for New York Life (the insurance company). If we follow these banners, we find ads directed at overseas Tamils. The bank ad boasts of "NRI Services," or services for nonresident Indians—that is, persons born in India, but now residing temporarily or permanently in other countries. The bank site focuses on the ease of completing wire transfers via the Internet, an important service for nonresidents from all locations. The New York Life ad bears the notice "Attention U.S. Residents Only," but is clearly targeting NRIs in the United States as potential clients. The AOL ads appear virtually anywhere on the Web, but this ad is aimed presumably at those who identify with the "America" as a macroscale context of communication rather than with India or Tamil Nadu. The ads accessible from www .dinakaran.com suggest a range of social contexts in the United States and India, with computer users moving between them. For these overseas users, the computer provides a means of accessing an otherwise inaccessible social space. These persons are highly privileged in comparison to most Indian citizens, yet the overall result of their virtual gathering on the Internet is to promote cultural fusions and hybridity that destabilize existing hierarchies (such as caste and patriarchy) as much as their gathering solidifies them.

To leave this page, we first click on the button that takes us into the English version, but discover that this page deals only with the Tamil film industry, so we back up and click "ENTER" to move into the Tamil version of the site, which has much more content.

4–5.

Even here we are not completely lost if we do not read Tamil. Roughly one-third of the site consists of the Tamil script or Tamil words written in the English alphabet; two-thirds of the site is in English. Context points to several explanations for the use of English on this page, chief among them being British colonialism, the dominance of English on the Internet, and late-twentieth-century U.S. economic hegemony in the global system. Many students in Tamil Nadu are educated in English from elementary school on, in particular those whose parents plan for them to go on to college.[5] The use of English as the lingua franca in India maps onto class and caste relations to produce a tendency for those Tamils who are successful (or who plan to be successful) to be more comfortable reading and writing in English than

5. Personal conversation with Sriram K. Raman, June 19, 2003.

in Tamil. As context, the Web communicates in ways that reaffirm upwardly mobile, globally extensible identities.

This page consists mainly of links arranged down the left side and across the top. These links are to news sites and cultural sites for Tamils. At the top of the lefthand list are five online Tamil news sources, followed by links covering cinema, health, an online novel called *Vanatham,* spiritual(ity), and sports. Below this list are the "Weekly Features": Arudum (astrology), Arokya (fitness), Ladies Park, and Rasi Palan (weekly fortune-telling column). Below these features are member services, including matrimonial listings (people seeking marriage partners, though not necessarily for themselves), mail, classified advertising, information about Chennai City, a monthly calendar, images of Tamil film actresses, and Tamil music. A bar across the upper part of the screen also offers various links: Vasakar Nenjam (feedback), Sinthanai Arangam (message board), Feedback, Troubleshooting, Chat, Links, News Search, Forum, and Chennai Metro.

This is a specialized hub site—that is, a site that includes a collection of links that hook up to various interests and purposes off the site as well as to other services such as discussion forums related to a certain affinity or interests. Such "node" sites (to employ Lynch's [1960] terminology of urban perception) are like a busy intersection that people pass through going different directions. Network theory has found that such hub nodes serve a special function in many different networks; they greatly reduce the separation between other nodes, which in our Web voyage translates into a saving of time. In topological diagrams, such sites appear as starlike formations with many links leading inward and outward (Adamic 1999).[6]

English and Tamil are mixed in a casual fashion in this site depending on the nature of the content, and Tamil words are written in either the Roman alphabet or the Tamil alphabet. Tamil and English are woven together in a systematic and structured fashion, demonstrating the linguistic hybridity common in postcolonial societies, akin to the use of

---

6. See also Martin Dodge's wonderful Web site, "An Atlas of Cyberspaces," online at: www.cybergeography.org/atlas/atlas.html.

"Spanglish" in the U.S.–Mexico border region or "Franglais" in Quebec and adjacent parts of Canada (Chambers 1995, 10, 163–67). English, as the language of science, medicine, and business, is essential to success in Tamil society, even if the Tamil language remains important in the home and as an element of Tamil ethnic identity.

The site caters to a particular ethnolinguistic group, but (like everything else on the Internet) it can be accessed from virtually anywhere as long as one has a computer and a phone line. In the most obvious way, it helps detach ethnic identity from location: anywhere there is a computer, one can be "in" Tamil culture. More subtly, it reattaches identity to location. Insofar as ethnicity implies a place of origin, a homeland, and often a set of particularly significant sacred places, any nodal site based on ethnicity implies a self rooted ultimately in place. An ethnic homeland is difficult to forget if one has a communication context in which to "reconnect" with this homeland through news and virtual gathering, even while one is temporarily working or studying abroad. Cultural separation from the homeland can be reduced even as the body is mobilized by modern ethnoscapes (Appadurai 1996). This form of extensibility reaches out from the placeless here and now to a there and then rooted in place; it establishes a foundation for the self in a place of personal or group origin rather than a foundation in personal skills, knowledge, and interests or in other nonplace attributes.

The use of the Internet to preserve ethnic identity is paradoxical in two ways: first, a placeless context is entrained to the task of perpetuating a sense of being rooted in place; second, a new medium is entrained to reaffirm an old sense of ethnic belonging. For individual users, communication via the Internet works like any other form of communicative action: it draws and redraws the boundaries of the self in space-time. All that is needed to bring together past and present, Tamil Nadu and Toledo, is access to an online computer with a Web browser. Yet this picture may still be too simplistic.

A rare ethnographic study of Internet users has found that "People on the Net have a greater tendency to base their feelings of closeness on the basis of shared interests rather than on the basis of shared social characteristics such as gender and socio-economic status"; it also found an increase in individualistic behavior: "Computer mediated communication accelerates the ways in which people operate at the centers of partial, personal communities, switching rapidly and frequently between groups of ties" (Wellman and Gulia 1999, 186, 188). Together, these findings suggest that online spaces constituted around ethnicity or race may fragment, specialize, or otherwise evolve toward a form based around shared interests. Their users may indulge in ethnic identity as only one of several or many online and offline constructions of personal identity. Insofar as this form of communication preserves ethnic identity and overcomes isolation while providing an impetus that resists parochialism and ethnocentrism, we have cause to celebrate this "habitation" of the Internet.

Whereas being Tamil previously depended on being born and living in Tamil Nadu or in a few other Tamil settlement areas in southern and southeastern Asia, it now entails involvement in either place or non-place contexts of communication, which facilitates community building without propinquity. As one adopts this mode of being in the world, the opportunities for "tasting" other communities and ways of being are multiplied. Arjun Appadurai asks, "What is the nature of locality as a lived experience in a globalized, deterritorialized world?" and answers enigmatically: "Culture does imply difference, but the differences now are no longer taxonomic; they are interactive and refractive" (1996, 52, 60), a comment that gestures toward Sherry Turkle's concept of "cycling through" identities that one enters episodically rather than permanently (1995, 12–14). The opportunities to appreciate and trust others, even those labeled as "different," must surely be multiplied when one ceases to "belong" to one community as a permanent inhabitant and instead passes or cycles through identities linked to various communities. The technology of the Internet does not dictate or determine this shift in personal identity, but to the degree it overcomes spatial and temporal limits that previously inhibited such cycling of identity, the Internet is a favorable context for this constitution of self. One can move thousands of miles without surrendering routines of involvement in a homeland language community and its other cultural facets; one can stay in one place and engage in an identity formed around places thousands of miles apart. Thus, being Tamil takes on an altered meaning—less categorical or automatic, but perhaps no less powerful.

The use of the Internet as a defense against racial contamination is an example of this phenomenon that rewards closer examination. Moving to the "Links" page, we therefore pick one link at random in the "Classifieds/Matrimonial" section of www.dinakaran.com. We find ourselves at a "leaf node" of a large Web site.

6.

Our location now is www.sysindia.com/matrimony/index.html. The simple page consists mainly of links to twelve matrimonial services, all managed by Bharatmatrimony.com, a company that lists more than half a million matrimonial advertisements. Each of the twelve services caters to a different ethnic community in India and its corresponding diaspora. Oddly, although we are now narrowing in on a specialized kind of Internet service, moving from a leaf node to a hub node, our path takes us through one of the leaf nodes of a different hub site. It is a bit like going from Los Angeles to Memphis through a suburb of New York City. But the Internet, as odd as its "geography" may be, still serves a basic function of supporting community formation, not unlike a bounded area of the Earth surface.

In the dark blue sidebar, we see links to "News, Discussion Forum, Kitchen Section, Greeting Card, E-Magazine, Immigration, Real Estate, Chat, Tamil Calendar, NRI Corner, Travel, Kolam, Baby Names, Reminder, STD Code, India Pin Code, and Links." These links let us know we are in another hub site, but if we follow links at random, we often find information on a single company. "NRI Corner," for example, takes us to the law firm R.S.N. and Associates, specializing in "Tax, Property, Personal Finance, and Investment Management." The "Immigration" link takes us to another law firm, "Smith, White, Sharma and Halpern," which offers free advice to those wishing to immigrate to the United States and probably acquires numerous clients through this service. The current page, "Matrimonial," is dedicated to a single online matrimonial listing service, although many online Tamil matrimonial services could have been listed. Therefore, what bills itself as "The first and leading Tamil Portal," is in fact largely a consortium of businesses that has created a corral-like virtual space in which to herd and concentrate potential customers (an increasingly common situation on the Internet). This communication context ensures a steady flow of customers to the affiliated businesses, and these encircled businesses in turn help sponsor the site and make it profitable. The macrocontext of the global economic system, economic competition in particular, has arguably shaped the microcontext of the Web site to the point that it reflects an intersection of class with ethnicity and other sources of social differentiation.

Accordingly, it is at this junction we can cross between potential user groups and follow a link that would be of interest if we were of a different, non-Tamil Indian ethnicity. A scroll bar allows us to pick our ethnolinguistic affiliation: Bengali, Gujarati, Hindi, Kannada, Malayalee, Marathi, Punjabi, Sindhi, Tamil, or Telugu. At random, we select Marathi. Segregation by company rather than by language or ethnicity has provided an "exit" from the loosely bounded Tamil sector of the Web.

7.

Marathimatrimony.com is a virtual meeting place for marriageable members of the Marathi community inside and outside of India. Like other Indian matrimonial sites, it is used by individuals looking for a partner as well as by family members arranging a marriage for a son, daughter, or sibling. This latter situation is an example of a traditional cultural trait taking up residence in a new technological context. Arranged marriage is a tradition in India, but prior to the computer, paper media were the norm for exploring listings, and before that such messages traveled on face-to-face social networks. Some of the potential brides and grooms live in India, but a fair number reside outside India, in the United States or elsewhere. This site advertises that it has more than fifty-three thousand matrimonial listings, indicating that even in a relatively poor country such as India many people are appropriating the computer and the Internet.[7] The users of this site construct themselves and others in ways that reaffirm traditional power relations based on caste, ethnicity, socioeconomic status, age, and, of course, gender. Following links within this site, we can explore a few listings to see how people adapt matrimonial content to this particular virtual context (all entries are copied without modification).[8]

A. This is the profile of my daughter Aarti. Her Date of Birth: 21.03.1973, Blood Group: O Rh+ve, Appearance: fair complexion, beautiful, slim, long hairs, matured, pleasing manners; purely vegetarian, religious minded, computer literate, two wheeler driving license. Languages: Mother tongue: Marathi & Hindi, English, Sanskrit are known languages, understands Bengali also. Education: M.A., (Pub. Adm.) LL.M., registered for Ph.D. Status: Practicing Advocate, contributory Lecturer. Horoscope: Kanya Rashi, Chitra Nakshatra, dwitiya charan, Madhya Nadi, kashyap gotra. Interest & Hobbies: Reading, Drawing, Music (Indian & Western), Traveling, highly social, swimming. Expected Groom: From educated cultured family, highly educated/Post Graduate, Professionals, Executives, aged up to 34/35 years, height nearly 165 cms, fair complexion, good looking, handsome, purely vegetarian, teetotaler, expecting no dowry, caste- no bar (preferably Maharashtrians), high caste Hindu: Brahmin, vaishya, Kashtriya preffered. Please send your reply with Bio-Data, horoscope and photo.

B. She is a brilliant student all these yrs. and has studied in convent school in Mumbai. Her parents are settled in Mumbai. She is a pleasant girl and belongs to a higher middle class family. She has been in USA for past 2 yrs and will be completing her PhD in April 2004. She possesses student visa and is likely to visit Mumbai around year end—2003. She prefers to continue to stay in USA even after her Phd. She is looking for a boy from Brahmin family from Maharashtra, tall, minimum MS, fair, and from a respectable family.

C. I am easy going, cultured and well-educated girl. Working as a

---

7. This estimate is based on the declared number of listings at the top of the site www.marathimatrimony.com.

8. The quotes are from online files at www.marathimatrimony.com, entries R157959, R145829, R158983, and R127861, respectively.

faculty member in a renowned computer institute. I work [also] as an lecturer in a science collage on temparary basis. Willing to do job after marriage if my new family members permit me to do so. I don't have much expectations a such. I am looking for well qualified, loving, caring, easy going life partner.

D. HI! I am a broad minded girl who is not very traditional but I am religious. I am a nature lover, I love travelling, visiting places & watching cricket. I am looking out for some one who will be like a BEST FRIEND to me. Someone who is understanding & caring, Who wants to settle in India. . . . .basically looking out for someone MADE IN INDIA WHO WANTS TO STAY IN INDIA.

Comparing these four entries (which are supplemented by a form on which applicants enter data in various categories), we immediately see two rather different forms of appropriation of this virtual space, one by the family of the marriageable person (A and B), the other by the marriageable person himself or herself (C and D). In either case, the computer serves as a surveillance mechanism to narrow the matrimonial gaze to "important" qualities of the eligible persons and to obtain a systematic, quasi-rational match. The new technology perpetuates traditional means of rationalizing the matrimonial search process, including the imposition of minute categories of caste and horoscope. To file a family member's listing (as in A and B) extends cultural traditions of arranged marriage. The same technology can facilitate a break with tradition, as the women who file their own listing (C and D) demonstrate. By doing so, they are on the assertive end of the Indian female-role spectrum. Thus, "Willing to do job after marriage if my new family members permit me to do so" is inflected by its context. Reading this comment in the context of American or European standards of gender equality would indicate a bizarrely submissive attitude, whereas interpreting the same ad in the Indian context (or rather contexts) would typically highlight the fact that this woman wrote her own ad, which marks her as fairly assertive (certainly less submissive than she sounds from within a North American or European cultural context). Although the emphasis in matrimonial ad "D" on personal interests and the need for friendship in the marriage would signal nothing out of the ordinary in an American classified ad, it indicates an exceptionally assertive and in-dependent-minded woman in its intended (Indian cultural) context, particularly in combination with the first-person mode of address. Iron-ically, this woman adopts a geographically conservative stance, whereas individuals and families living abroad and presumably more thoroughly exposed to non-Indian culture pursue the more socially conserva-tive stances. Any communicative action must therefore be placed in the proper cultural context in order for it to make sense, even if it is accessible from a vast range of physical contexts owing to the use of the Internet.

Listings A and B demonstrate that a twenty-first-century technol-ogy can easily be adapted to a conservative construction of femininity. Here the woman is placed on display in virtual space, and technology is used to demonstrate everything of "relevance" in that rationality: blood type, astrological signs, and credentials from computers to "two-wheelers." The woman has been objectified or commodified through the use of telecommunications, but this objectification has deeper roots than capitalism or electronic technology. Indian (Maharashtrian) patri-archy is a driving element of the context surrounding the matrimonial ads, and C and D show how extensible agents negotiate with or openly challenge these patriarchal power relations in a virtual space. Again, we see a new medium opening up opportunities for individual agency, de-spite that agency's susceptibility to the disembedding and distanciation of coercive social structures. For some of this Web site's users, love is part of the equation, and extensibility is under self-direction, whereas for others love is irrelevant or of minor importance, and the self is not supposed to develop its own extensibility.

In short, a single communication context—a single Web site with a single administrating body—can support various and conflicting con-structions of the extensible individual. On the one hand, the Internet serves as a way to preserve family status by perpetuating values of In-dian cultures in a condition of diaspora. Classifying and displaying a prospective marriage partner's marketable skills, potentials, and promise within a culturally determined grid of suitability produce an alliance of one family with another (despite the fact that the mobility motivating the use of the Internet means that the families may come into contact only rarely and that the members of the couple will be most dependent on each other, whether or not they fall in love). On the other hand, the Internet serves as a way for a single person to perform his or her identity for prospective partners, who are expected in turn to perform their identities, with the primary goal being the mutual discovery of a uniquely compatible individual, a soul mate somewhere on the planet. In no way does the Internet favor one of these models over the other. Indeed, it makes possible the preservation of a traditional "family-centered" model of marriage despite the fragmentation of communities caused by migration to the United States, where this tradition is virtually nonexistent. Likewise, it makes possible the adoption of the individualized "soul mate" or "love marriage" model wherever one has access to a computer, even if this place is a conservative part of India where the family-centered model normally rules. The technology circumvents place as a factor determining cultural behavior and socialization, though places remain and are not annihilated.

More can be discovered if we back out to a site that acts as a point of entry to www.marathimatrimony and thirteen other ethnically defined Indian matrimonial services.

8–9.

Bharat means India, and from Bharatmatrimony.com we can access matrimonial Web sites tailored to each of India's major ethnic communities: Assamese, Bengali, Gujarati, Hindi, Kannada, Kerala, Marathi, Marwadi, Oriya, Parsi, Punjabi, Sindhi, Tamil, and Telugu. The popularity of online matrimonial listings is clear. Six Indian matrimonial sites, located in a brief search on June 15, 2003, collectively claimed more than 1.3 million members. Does the "high-stakes" communication situation surrounding marriage encourage the use of high technology in India and among NRIs? The critical theory equation "power equals knowledge" serves us well: as a symbol of knowledge, the Internet seems to have the power (that is, it *symbolizes* the power) to filter out the less-desirable matches. Although Americans trust this filtering to the dating process, many Indians see dating as immoral and therefore prefer to trust astrology, technology, and one or more of the deities in the Hindu pantheon.

Bharatmatrimony.com has far more inward links (15,269) than outward links, indicating that for many users it is a destination rather than a means of reaching other sites. But we can find a way out of most online cul-de-sacs if we look hard enough, and this one is no exception. A sidebar contains links to various marriage-related services as well as information on starting a franchise with the matrimonial service. We follow the "E-Greetings" link, which leads to a page with various e-mail-based greeting "cards." From here, we take another link to a similar service on Yahoo India.

10.

The e-greetings on Yahoo India are a mixed bag—culturally, aesthetically, and technically. American and Indian, Christian and Hindu celebrations and observances are listed together. Some cards contain still images that are essentially the same as printed greeting cards and therefore reflect a minimal adaptation to the new communication context, whereas others contain animated digital "gif" images (short movie clips), indicating a use that begins to respond to the potentials of the online medium. The principal Indian holidays and festivals are not apparent on the first page; we must click on "Holidays," then work our way past April Fool's Day, Groundhog Day, Passover, and Easter before we get to an alphabetical list of Indian holidays. The 44 Durga Puja cards and 278 Diwali cards are severely outnumbered by the 699 virtual Christmas cards. The visual and technical quality of the animated Christmas cards greatly surpasses the quality of the few animated Indian holiday cards. A religious hierarchy placing Christianity above Hinduism in Yahoo India is difficult to miss. It appears that Herbert Schiller is justified in arguing that "The free flow of information, as implemented, has meant the ascendance of U.S. cultural product worldwide" (1995, 19). But the cause here lies not with the free flow of information, but rather with the economic, cultural, and institutional contexts of

Yahoo India. To see if this failed adaptation of an online service provider to Indian culture is typical of the rest of the Yahoo India Web site we back out to "Yahoo India" (in Yahoo.com) in the upper left corner of the page.

11–17.

Yahoo India appears to be a good candidate if we are headed back to the United States and the University of Texas because its close ties to (the U.S.) Yahoo should make it an easy route to the major sites in the United States. In fact, the University of Texas at Austin is only three links away at this point (using the search engine of Yahoo or Yahoo India), but we will take a more circuitous route home in order to compare Yahoo India with GHEN's services.

Yahoo India illustrates the process of centralization on the World Wide Web. In 2003, this was not a tremendously popular site. Reached by only 1,774 links, Yahoo India was less than one-twenty-fifth as popular as www.hindunet.org, one of the GHEN sites, which could be reached by more than 51,000 links.[9] Yahoo India (by late 2004) had mushroomed in popularity and had seventeen times as many links leading to it as did www.hindunet.org. In some cases, Yahoo India adapts well to its intended audience: the movie reviews cover Bollywood (Bombay film industry) news, not Hollywood, and "cricket ratings" earns its own special link on the main Yahoo India page. But a closer look demonstrates at best a truncated adaptation to social context.

If we click on "Society and Culture" (12) and then on "Religion and Spirituality" (13), we would expect to find quite a bit of information about Hinduism as well as significant numbers of links relating to Islam, Jainism, Buddhism, and Sikhism. This is not always the case. To see this we can follow the "Pilgrimages" (14) link. We discover instead that of the eighteen links on this page, fourteen (78 percent) connect to infor-

9. The link counts were calculated on the AltaVista search engine using the special command "link:URL." Yahoo India was about one-one-thousandth as popular as Yahoo.com, which could be reached by more than 1.5 million links.

mation about Jewish or Christian pilgrimages. Of the remaining four links, two are interfaith links with information on pilgrimages in various religions, and only two are specifically about Hindu pilgrimages (though many sites about Hindu pilgrimage do exist on the Web). Missing from Yahoo India (as of July 2003) are links relating to the major pilgrimage centers at Prayag (Allahabad), Haridwar, Ujjain, Nasik, Benares (Varanasi), Vrindavan, Kedarnath, Amarnath, Badrinath, or even the Maha Kumbh Mela festival held every twelve years at Prayag, which attracts more than 25 million people (and therefore constitutes the largest religious gathering in the world). The paucity of information about important Hindu pilgrimages on Yahoo India cannot be explained by factors intrinsic to Hinduism. When an American company such as Yahoo.com creates a local subsidiary such as Yahoo Web Services India Private Limited of Mumbai, the economic relationship is an element of context that distorts sensitivity to local culture. The character of this Web page suggests that Yahoo's specialized services such as Yahoo France, Yahoo Brazil, and Yahoo China may also fail to adapt to the "local" culture. Schiller would interpret this failure as cultural hegemony, and the growth in popularity of such sites suggests they may "take over" and squeeze out truly "local" hubs although they appear to fall short of what local Internet users would expect of an Internet hub.

Despite these arguments, some locally relevant links are fairly well represented in Yahoo India. If we follow the links to religious "Holidays and Observances" (17), for example (passing through "Hinduism" [15], then through "Faiths and Practices" [16]), we come to a much more complete page from a Hindu perspective. Twenty-five holidays are listed as headings for new link pages, and nine pages with Hindu holidays are listed individually. Although this list is unusually complete for Yahoo India, if we follow the top link, "Festivals of Bharat," we can find an even more comprehensive listing on a GHEN site.

18.

We have arrived at www.hindunet.org/festivals. This is in fact only one page on the labyrinthine Hindunet site, with its hundreds of interlinked pages and the even more extensive network of interlinked GHEN resources. Here we find sixteen Indian festivals, with links to detailed explanations beside sixteen small images in the Indian aesthetic tradition. This site is clearly by and for Hindus rather than a commercial site seeking Hindu traffic. The page is rather complicated in appearance, with sidebars and buttons across the top connecting to other parts of the large and popular Hindu Universe site. If we follow any one of the links on this page to a particular holiday (all Hindu), we find in-depth information on the holiday, several pages in length with illustrations. There are recipes for foods associated with the holiday, schedules of holiday events in different places, links relating to the holiday and associated religious information, and so on. In addition, a link also leads to holiday goods that can be purchased online, a detail that suggests many users of the site are outside of India. Hindunet.org greatly surpasses the Yahoo India site in cultural specificity, blending the local (culture) with the global (technology) more effectively. Globalization does not necessarily imply homogenization.

19–20.

If we back out to Hindunet.org, we can find another application of the Internet by and for Hinduism that has very little to do with Western concepts of "enlightenment" and "rationality." This appropriation of technology also does not in any obvious way reinforce the global capitalist economy. Quite the opposite, it appears to offer an alternative to scientific causality and instrumental rationality. By choosing "Puja and Samskars" (prayers, rituals, sacraments) on the Hindunet.org site, we arrive at a collection of excerpts from a book promoted on an affiliated site, Hindubooks.org (a GHEN Web page). A small taste is the following:

> Achyutaya namah, Anantaya namah, Govindaya namah . . . Achamanam, that is, sipping of the water with the above mantras, one sipping for each mantra, will remove all the ills of the body—and mind. This is called Namatrayividya or the worship with the three names,

which will cure all diseases, physical and mental. There is a well-known sloka to the effect that the medicine constituting the repetition of the three names of the Lord Achyuta, Ananta and Govinda will certainly cure all diseases.[10]

It bears repetition that there is no inherent logic or bias to the Internet. The context is subject to individual and shared agency. Its content can be Christian or Hindu, materialist or spiritual, rational or mythical/magical. Despite the quantitative dominance of the Internet by the United States (e.g., the most servers, the most sites, the most common Internet language), the content of the Internet is not straight-jacketed by Western rationality or its American expression, and users will seek out what is most relevant and meaningful to them, particularly because every site is equally close (for all intents and purposes) in time-space.[11]

21.

On the Hindubooks site, www.hindubooks.org, clicking on "magazines," "publishers," "Hindu Software," or "Bookstores," takes one to Hindulinks.org, another site managed by GHEN. This site is a hub site for the "corner" of the Internet related in any way to Hinduism and related belief systems. The page advertises "29,263 HinduLinks for you to choose from." The categories are: "Arts; Children and Youth; Customs; Dharma [religious way of life] and Philosophy; Frequently Asked Questions; General Sites; God, Sages, and Gurus; Hindu Merchants; Hindus around the World; History; India; Interfaith Relations; Internet Books and Resources; Jain Dharma; Languages; Sciences; Scriptures; Seva; Sikh Dharma; Social and Contemporary Issues; Temples, Yatra [pilgrimage] and Organizations; and Worship." Sidebars provide internal links to "Hindu Resources, Kids and Youth, Women, Health and Life Style, Marriage and Family, Pujas and Samskars, Spiritual Tourism,

10. Online file at: www.hindubooks.org/sandhyavandanam/achamanam/index.htm.
11. For measures and maps of Western dominance on the Internet, see Balnaves, Donald, and Hemelryk Donald 2001, 82–91.

Community, India, Hindu Shops, Multimedia, and Education," as well as user-created sites and introductions to Hindu, Jain, Buddhist, and Sikh dharma.

A sidebar on the right offers links to ads for books, music recordings, videos, magazines, wireless phone service, and airline tickets. These intrusions of the market do not seem much like colonization of the lifeworld because the products reinforce a belief system and a way of life that is separate from and in many ways opposed to "the system" of rational action coordination identified by Habermas. As a context of communication, this hub more fully accommodates the intersubjective and subjective elements of Indian cultures in comparison to the Yahoo India hub, asserting a set of values that resists Western rationality (of the kind that Habermas tries to restrict as well as of the kind he endorses). More than 2,000 links are provided to "Gods, Sages, and Gurus," for example. More than 4,700 links are provided to "Temples, Yatra, and Organizations."

Thus far we have not approached anything like Habermas's ideal speech situation because the sites have not supported interactive communications such as discussion and debate, but Hindulinks.org begins to fulfill this role. Visitors can join in on discussions ranging from religion to politics to women's rights to defense. These discussions are open to anyone who cares to join in. Now the "proximity" in virtual space to non-Western, nonrational thought (Eastern religion) appears to serve a role Habermas did not predict. It helps establish a "space" for non-Western debate, and this space potentially supports the consensus-reaching form of discourse that Habermas associates with rationality. Registration is required, but registrants can leave most fields in the form blank, meeting Habermas's first criterion of an ideal speech situation— namely, "every subject with the competence to speak and act is allowed to take part in the discourse" (1990, 89). The physical separation of discussion participants removes the threat of violence, thereby negating place-based inequalities that would inhibit discussion. The combined ease of access, global reach, and impossibility of physical violence create a close approximation of Habermas's ideal speech situation. Let us therefore look closely at the exchanges "contained" in this virtual space.

The category with the most postings was "General Hindu Discussion" with 6,798 postings. In "Relationship between Hindus and Moslems [*sic*]" (3,844 postings as of June 16, 2003), a three-way exchange took place between a Hindu, a Muslim, and a Christian in late February and early March 2003 that bears quoting at length (all entries are copied verbatim).[12] TM11 writes under the heading "Conversion Law in Gujrat/India":

> I am confused if most strict Hindus believe "all paths lead to same truth and all religions are OK," then why have a law against conversion?
>
> Can anyone explain this?

The question refers to the authority of the Gujarat state government to prohibit organized religious conversions. The bill was advanced shortly after the Hindu nationalist party, the Bharatiya Janata Party (BJP), took power in Gujarat. It was similar to laws in five other Indian states to stop Buddhist and Christian conversions of poor, lower-caste Hindu (Dalit) communities, to whom they would also offer aid. The BJP perceived the aid as an unfair inducement to conversion. Fines and jail terms were therefore imposed on non-Hindus who employed coercion or "fraud," meaning any form of material assistance, to promote their religion. A discussion participant with a clear religious and locational identity, Americanhindu, defends this policy:

> [Y]ou do agree there are several pathways to the same goal. Right?
>
> Why do you have to tell someone that someone must follow only certain path? Let them chose.

12. The entire exchange was found in the forum section, under the heading "Hindutva and Secularism," and the subheading "Relationship Between Hindus and Moslems [*sic*]." Online file read June 15, 2003, at: www.hindunet.com/forum/showflat.php?Cat=&Board=moslems&Number=27653&page=0&view=collapsed&sb=5&o=&fpart=1.

Hinduism is against "conversions" to all other religions mainly because (1) It has everything, meaning several pathways to chose from; (2) If forced or coerced or bribed, people do not really connect w real GOD; (3) Freedom of choice is the main aspect of Hinduism; (4) It is about Diversity; (5) yet unity w diversified pathways.

There is only one Big UMBRELLA here and it is called Hinduism, under which all other religions take cover.

I hope this helps a bit.

What religion are you following and why do you think conversion is right or okay? How would it help you?

I love hinduism the best.

TM11 responds:

Is it right for a government/any government to tell a person what he/she can/cannot believe? ESPECIALLY what religion he/she can/cannot convert to ????

Regarding "forced" conversion, there is no known case that exists in the last 50 years that I am aware of. . . . If a person changes because of money/food/clothing, then what right is it of the government to intervene???? If the person is NOT sincere in his/her commitment, is it not between that person and his/her Creator?? Certainly not between him/her and his/her government. . . .

Can you image such a law being passed in America in this day and age????

This law is also very very ambiguous since it mentions "propoganda"? What is that supposed to mean??

Regarding "my religion," I choose to become muslim from Hinduism. My family is Hindu and I studied Islam in college where I made the decision to become muslim.

It was not because of "hating" other religions or being "forced" into it. Islam in arabic means peace and submission to God and God alone; not to any white man or indian man or idol or chineese man. . . . There are absolutely no idols or images of God in Islam. . . . And worship is directly to God, required 5 times a day, not through a white-man or intermediatary. . . . Worship/praise belonging to the One God alone, upon Whom all depend, Who is nobodies child, Who has no children, Whom nobody can compare unto. . . .

Think about all the people who became muslim in the last 50 years: so many famous Indians have become muslim, including A.R. Rahman. . . . Do you think somebody could have force Mike Tyson or Malcom X???? Maybe it was the clear concept that asks How can the created become the Creator? Is there any comparison between the created and the Creator? Worship/forgiveness/salvation comes from God and God alone in Islam.

Now I ask you again: Is it right for any government to choose what religion you can/cannot convert to???

Another participant, jesuslovesyou, enters the exchange at this point, accusing TM11 of hypocrisy, arguing that conversion is punishable by death in every Islamic country, and interpreting India's anticonversion laws as a necessary means to control forcible conversion to Islam:

Before answering, I want you to consider a hypothetical situation. Suppose a Muslim in Pakistan wants to become a Hindu, what would be his fate? He would be stoned to death. The same thing will happen in any islamic country, because it encourages fanaticism. Please do not lie that there is religious freedom in Muslim countries, the whole world knows the truth that Islam treats other religions as second-rate. So I suggest you ask the same question as to why Muslims ban religious conversions??? Are you saying that muslims have the right to ban religious conversions, and Hindus don't have the same right???

Second, you are confusing forcible conversion with voluntary conversion. For instance, I believe in jesus' teachings, though I am not an orthodox christian. Suppose i speak to you about Jesus' sermon on the mount and you also love it and start believing in it. This is a clear example of voluntary conversion, and the government allows it. What the Gujarat govt. and TN [Tamil Nadu] govt. doesn't allow is forcible conversion, like they do in North East. The best examples are Nagaland and Manipur, where if you don't convert, sometimes even your arms and legs are chopped off. Jesus taught peace and love, so why don't you ask missioanaries why they resort to violence when people refuse to convert???

Another question for you. You claim to be a muslim, so why don't you ask why Pakis, Banglasehis, Saudis and almost entire Muslim world rgularly persecute Kafirs, especially Jews and Hindus??? At least in India, you have enough freedom to question the govt., conversion laws etc., but had you asked the same question of Pakis, they would have stoned you ASAP. So much for tolerance!

It is amazng how you ask these questions only of hindus, and you don't have the guts to ask muslims, because you know that Muslims wouldn't consider your question at all. You will be branded an infidel and stoned. So first try to understand why Islam hates other religions, and then you can worry about Gujarat's so-called fascist policies.

The harvest truly is plenteous, but the laborers are few!

Another participant—whose user name, Wesley Kulshresh, sheds little light on his identity—enters the discussion at this point, agreeing in principle with jesuslovesyou's assessment of Islam, but shifting the debate toward a more confrontational tone and reasserting the status of Hinduism over Christianity and Islam.

YES, it is natural in defense for the Hindu government to set limits to protect what BHARAT stands for. . . . this was the wholee reason of the SouthAsia splitting up. . . . WHICH MANY HINDUS WERE NEEDLESSLY KILLED IN THE PROCESS

HINDUSTAN IS FOR HINDUS. . . . and yes they have every reason in the world to protect their heritage and beliefs, Hindu Philosophy is eternally tolerant, but the Hindu *ungodly* people will also control the posessed Allah-cocksucking-feared-Muslims, whom yet falsely use their Allah's name before every list of their ignorant *evil*, senseless wishes as an sorry excuse of their' religion. . . .

India is too tolerant to their native minority of Muslims whom indirectly truely say . . . "You are not a follower of Allah, therefore you are going to burn in hell anyway, so [censored] you". . . . while the Muslims of Pukestan run or kill all non-believers out, and even theirself wish to move out've their Allah-ass-whiped currupted country they created, to the South into India.

The Internet facilitates the trading of invective to the degree that the term *flaming* has been coined to signify this hateful rhetoric. By matching the elements of Habermas's ideal speech situation—universal access, unlimited scope of debate, freedom from coercion—the Internet encourages flaming as much as debate. Wesley Kulshresh's anti-Islamic passion does not build any trust or love for jesuslovesyou, however. Wesley continues:

Islam, Christianity, and Roman have one belief in common which is . . . THE ALMIGHTY GOD CREATED US OUT'VE NOTHING AND WE WILL BURN FOR ETERNITY IF WE FAIL TO FOLLOW HIM . . . while a punkass Muslim or crazed Christian will say to a helpless Hindu . . . "Convert or go to hell". . . . a strong Hindu/SIKH would reply . . . "As a Hindu, I sincerely accept and enhance who you are, but don't [censored] with me or try to change me because I AM HINDU . . . first understand what it means to be HINDU.. otherwise, I do not neeD YouR KAma N sheit like dhat~!!!"

. . . .

Truth lays withen the eye of the beholder; much Love and Respect for all

The last six words of this tirade are clearly disingenuous and therefore are easily challenged with regard to the validity claims of subjective statements. The earlier portions, as unpleasant as they may be, do not strictly depart from rationality if we interpret them as statements of subjective perception. A rational discussion tactic would be to interpret them as testaments to personal experience, which may or may not be sincere, but rather than take this approach, Americanhindu follows the turn toward flaming by directing his (or perhaps her) anger toward the Christian participant rather than toward the Muslim. Note that the language employed in this contribution begins to include more of the abbreviations peculiar to Internet communications, revealing an association, at least in this particular communicator's mind, between Internet "speech" and intolerance as an intersubjective stance. This intolerance appears after a delay of four paragraphs:

> you wrote:
> "you are confusing forcible conversion with voluntary conversion. For instance, I believe in jesus' teachings, though I am not an orthodox christian"
> This is what people of all other faiths resent. you cunningly try to infilterate and twist their religions giving various methodolgy or titles to it. Hindus and muslims or jewish people resent that totally. It does not matter how you do it. All conversions are wrong to begin w.

> Any one who volunteers to convert to another religion has to be 21 yrs old and above. anyone who wants to convert has to be very well informed of his own religion from his own family traditions first. Conversions are not jokes.

> People in those countries esp families have a right to establish a rule about conversions in the form of laws. No family or a child under 21 should be tricked in any manner to convert to other religions w/o family approval. All christians do is to split up the families, twist their family values or whatever they hold to be very precious. Enough of that now.

> you wrote:
> "Suppose i speak to you about Jesus' sermon on the mount and you also love it and start believing in it."
> you have no businees in the first place even to speak to those peo-

ple of other faiths about your stuff. you may want to speak to your own christian people first to make them better christians. They have to learn to leave other faiths alone and respect them for what they are.

Here Americanhindu's rhetoric shifts toward "flaming":

No one in the right mind would think your scriptures have any value to tellyou the truth. It is full of flaws, nothing good for any civilization. WAtch jerry springher show or manshow or loveline or howard stern show first. Tell those dudes about your sermon on the mount etc. Let me see their reaction! LOL:)

Why is that America falling behind in moral values now? What does jesus dude teach them? mary taught it is okay to be a slut and get a baby w/o her hubby knowing about it? mary had a child w a black man and jesus was a "malatto?". Did you know that jesus was Black man only? Hmm. I wonder, how kids would think if they know about this.

I love hinduism the best.

We can see enough in these passages to recognize the problematic nature of online discussion.

The Internet as a communication medium has the potential to bring together people of radically different perspectives and backgrounds who might not otherwise engage in discussion of any sort—to facilitate just, sincere, and truth-seeking communication by removing the possibility of physical violence or persecution of one conversation participant by another. This virtual space of social engagement could foster mutual understanding by making hegemonic and counterhegemonic discourses equally accessible (Warf and Grimes 1997; Castells 1999; Hénaff and Strong 2001; for the opposite tendencies, see Castells 1996, 1997, 1998, and Taylor 1997).

It is therefore all the more disturbing that when we "enter" these spaces, we often do not find conversations oriented toward mutual understanding. We must be clear that instrumental rationality has not corrupted the language used in the Hindulinks.org online forum. The problem seems to arise from precisely the opposite direction: claims are made that do not aspire to be interpreted as rational discourse and that do not reflect the characteristic deceptions (and clarifications) of this mode of argumentation. So the Frankfurt School tells us little about what is wrong. Habermas's defense of rationality serves us better here. Subjective and intersubjective validity claims are so hopelessly mixed up with subjective disputes in the opinionated exchange given here that no mutual understanding can possibly come from it. Name calling ("punkass muslim"), boasting ("one Big UMBRELLA"), insulting ("no one in the right mind would think your scriptures have any value"), and gross understatement ("I love Hinduism the best")—all carry the debate into a realm where rational thought founders, and it is precisely this irrationality that derails the search for common ground.

What is interesting from a Habermasian standpoint is that an Internet discussion forum is close to an ideal speech situation. All four participants can communicate freely, expressing their opinions without fear of being oppressed or coerced by the others. But it is precisely that absence of hierarchy, of social structure, of imposed social order backed by the threat of social sanctions that facilitates the rapid descent to flaming, exaggerating, and hateful name-calling. The Internet's vaunted anonymity also undermines the rationality of the dialog, but again the prime cause can be traced to a lack of social structure in this peculiar type of communication context.

This unexpected finding suggests that Habermas's model is incomplete because it overlooks the need for *a sense of social structure and shared destiny* in addition to the freedom to speak and the absence of coercion. The proposed addition to Habermas's ideal speech situation can be summarized as *elements of community,* and for these elements to exist in a distanciated context would require the presence of some degree of trust and some mutual love (in the antiquated sense) between participants. Furthermore, the distanciated context does not simply reduce people's reluctance to enter into an argument; it removes the communal framework that would preserve the creative labor of achieving consensus. The extensible individual's characteristics that I have sketched in chapter 2 are not merely descriptive in nature; they also indicate precisely those qualities that must be strengthened if people are to bring new contexts of communication such as Internet chat rooms and discussion forums more closely into line with the ideal image of democratic deliberation. More profoundly, we might surmise that without an extension of moral commitment into global social space—a sense of community or shared destiny—communication in distanciated forums, even those situations approximating Habermasian ideal speech situations, are extremely unlikely to generate mutual understanding. A "critical" sensibility regarding communication contexts, which seeks primarily to expose power relations, lacks entirely the vocabulary and pragmatic focus that would promote such a vision. Constructive creation, not critique, is needed.

Unfortunately, the Habermasian perspective suffers not only from the failure to appreciate the positive role of control, but also from the weakness that its proponents generally disregard elements of communication that lie outside conventional descriptions of (Western) rationality. These elements include the signals and symbols described in the previous chapter and many others not mentioned. Although a lack of rationality may often result in a breakdown of communication (as shown in the flame session), it serves purposes not appreciated by Habermas. Our journey therefore takes us through this kind of symbolic communication context before we head "home."

22.

Many of the links in Hindulinks.org lead to pages on FreeIndia.org (also under the management of GHEN), a site that serves as a resource for both Hinduism and Hindu nationalism. Picking a link at random on FreeIndia.org, we find an explanation of Ganesha, the elephant-headed god:

Ganesha's bulky head symbolizes his extraordinary intelligence. His ears are broad like winnowing pans. You know winnowing-pans are used to winnow grain. What happens then? The husk and the grain get separated. So does Ganesha distinguish between truth and untruth. It may also be said that the broad ears symbolize his capacity to listen to the prayers of all his devotees with great attention. While his ever-moving trunk teaches that one should be active always, his single tusk denotes single-mindedness in action. His huge belly signifies that the entire Brahmanda (universe) is hidden within Ganapati.[13]

The overall effect of the elephant-headed god as a signal of attentiveness, wisdom, activity, and single-mindedness is completely alien to a Western (Christian/Euro-American) worldview of elephants or of God. Both the imagery and the explanation of Ganesha can be transmitted over the Internet, in English, despite the Internet's foundation in "Western" technology. The distinctly non-Western symbolic explanation problematizes the critical theory critique of Western rationality as a conquering force that expands wherever it is introduced because although it is disseminated in English, via computer, this explanation of Ganesha retains its non-Western symbolism. We can separate the technological and social aspects of communication context and surmise that technologies derived from Western rationality are containers for a wide range of communication contents, not all of which promote Western rationality or Western strategic objectives. Outright resistance to Western strategic objectives is also apparent.

This resistance is evident in the "freedom fighters" section of FreeIndia.org, where one can read about the lives of nine "freedom fighters" who fought for India's independence. The section dedicated to the anti-British insurgent Khudiram Bose tells how Bose used the phrase "Vande Mataram" (I salute the Mother) and the image of Bharat Mata (Mother India) to rally opposition to British rule. When questioned about his decision to kill a British officer, Khudiram Bose explained his courage in terms drawing on this image of the nation as parent figure (symbol): "Bharat Mata is my father, mother and all. To give up my life for her is, I consider, an act of merit. My sole desire is only this. Till our country wins freedom, I will be born here again and again, and sacrifice my life."[14]

It would be absurd to consider this text an "accommodation to

13. Online file at: www.freeindia.org/biographies/gods/ganesha/page1.htm.

14. Online file at: www.freeindia.org/biographies/freedomfighters/khudiram bose/page10.htm, accessed June 15, 2003.

dominant currents" of Western thought (Horkheimer and Adorno [1944] 1972, xii), and this site puts the lie to Babington Macaulay's naïve claim that "No Hindu who has received an English education ever remains sincerely attached to his religion" (quoted in Smith 1963, 339, as cited in Anderson 1983, 91). It does work to promote the nation of India, which is an instance of nationhood on the Western model of an "imagined community" (Anderson 1983), but this nation is a container for communication that is not necessarily tending toward Western values or norms. Images on this site, such as the "unique Freedom Struggle Mini Movie," help glorify the struggle for autonomy, which is an empty symbol easily embellished with religious, economic, political, and cultural subtexts. The motherland imagery supports political opposition to external Western hegemony and the motif of reincarnation combines a religious symbol with a signal to resist foreign influence and domination. Resistance may be situated in the context of the web, using symbolism as content, but neither content nor context can be understood purely in the political framework.

Our journey has been long, and time is pressing. We now turn toward home. The surprising aspect of the Internet is that virtual places far apart in ideology may be rather close in the topology of nodes and links. Our onramp to the University of Texas, so to speak, is given a prominent position on the FreeIndia.org page, with its own box near the bottom left corner of the screen. The penultimate stop is www.vidya patha.com, which is billed on FreeIndia.org as "The largest Indian educational portal."

23–31.

Vidyapatha.com is nationalist in its own way, displaying "Student's Power Nation's Power" on a banner at the top, yet as instrumental rationality governs the choice of college education, users can easily find links to universities and colleges in Australia, Canada, Germany, the United Kingdom, and the United States. Here the model of colonization by Western ideology is more applicable, although as the matrimonial ads suggest, a college education in the United States does not necessarily lead to abandonment of racial prejudice against intermarriage or to the weakening of cultural elements such as particular constitutions of the individual. Clicking the "Education Abroad" link (24) and the "U.S.A." link (25) leads to an exhaustive list of links to U.S. institutions of higher education, where we can easily locate the University of Texas system (26). From there, we close in on my Web site by navigating to the flagship university (27) to the departments link (28), and by selecting "Geography" (29) from the list of "Colleges and Academic Units." Our final link appears on the Department of Geography faculty page (30). Clicking on my name takes us home to where we might start the whole circuit again with a single click (31).

If it seems as if our path through the online world has been long and tortuous, we need only call to mind the rigors of the actual trip to India, even in the age of jet travel. Non-Internet media are a poor substitute: finding a library outside of India with an exhaustive collection of up-to-date documents from and about India is quite difficult and would again involve travel for the vast majority of prospective "visitors." So we may raise our eyebrows at Howard Rheingold's (1993) claim that the Internet supports "virtual community" or William J. Mitchell's bold assertion that "we have reinvented the human habitat" (1995, 166), but the Internet's institutional-technological framework clearly supports ways of coming together and being together that are unprecedented and that presumably will lay the groundwork for new ways of relating and constituting the human self.

If Internet-mediated interaction seems prone to recapitulate narrow ethnocentric attitudes; to encourage racism, nationalism, or classism; and even to promote the subordination of personal desires to a new form of surveillance, we should keep in mind the technology's potential to be appropriated in ways that break down monopolies of power not only through political contestation, but also through symbolism, religious and otherwise (Adams 1996, 1997). For all the shortcomings in the current online forums, it is not unimaginable that they can be structured and adapted in ways that promote democratic civil discourse.

Above all, the Internet is a context of communication in which linear distance translates only slightly into time distance. Therefore, "places" such as Hindulinks.org provide Westerners and immigrants to the West with essentially instant access to communication situations that are clearly non-Western: the "connection" is not transparent (nothing is the same as being in India, immersed in those cultures), but it juxtaposes Western culture with a culture that is distant at the level of symbols, antagonistic at the level of signals, and alien or even confrontative at the level of signs. It has been argued that simply having access to an "ideal speech situation" is not enough to ensure the move toward a common ground of mutual understanding. I have shown that people must appropriate new media by developing the characteristics of themselves that are linked most closely to extensibility: trust, love (or at least mutual respect), altruism, and creativity. The alternative is the surrender of virtual spaces to capitalism, nationalism, racism, sexism, and religious chauvinism. The colonization of this space by instrumental rationality and the hegemony of state or capital are again only parts of the problem. The challenges to the extensible self in the Internet, or in any medium, are as complex and multilayered as the full range of physical, social, and communicational environments people inhabit. Every medium is a microcosm of the world inhabited by its users.

5

Communicating a Better World

> [T]he world *is* neither this nor that, but what we make it; and we
> need both this *and* that perspective before we can even begin to
> understand.
>
> —Helen Couclelis, "The Truth Seekers:
> Geographers in Search of the Human World"

THIS FINAL CHAPTER looks carefully at how we represent the world.
The ways we represent the world shape the ways we understand the
world, and the latter ways in turn affect the ways we interact with the
world (Buttimer 1982). This causal connection between representation
and action is one facet of the grounded nature of communication. It
forms part of a larger cycle because our actions lead to environmental
changes (crops grow or they fail, water is captured behind a dam or the
dam breaks, and so on), which create opportunities to reflect on what
has happened as a result of our actions (and indirectly as a result of our
communications). These opportunities initiate a new cycle of communicative action. Before we act, we state and debate alternative actions, although this deliberative activity sometimes occurs rather obliquely, and
the debate seldom proceeds in the absence of exclusion and social hierarchy. This then is the reflexive cycle of communicative action recognized by structuration theory and by Habermasian communicative
action. Consequently, communication is grounded in multiple ways: in
the sender's and receiver's subjective experience, in the intersubjective
relations of the social context(s) in which sender and receiver are able to
connect, and in the objective phenomena referred to and in fact symbolically modeled in the content of the communication.

This cycle of communicative action, material action, and environmental response is like a vast river with many small eddies. One such eddy is academic communication about the earth as the home of humankind—the field of study called geography. Within that eddy, there are subeddies, swirls of reflexive thought about topics as varied as agricultural land-use change, erosion, public transit, and landscape ecology. In part, this chapter is intended to consider the implications of the context in which geographical writing occurs—specifically, the large flow of action in the world and reflexive thought about the human-environment relationship. To do this, I first must visit another eddy—that of historiography, the study of ways of conducting historical research and writing. Historiography provides a model for the evaluation of geographic writing. This chapter also brings us back one last time to the idea of the extensible individual and considers the moral obligations and ethical responsibilities of this oddly stretched-out being.

Communication as Grounded Action

To communicate, as either sender or receiver, is to *act*. I mean this not just in the sense that we make a bit of noise or scratch on paper, but in the more important sense that we share ways of "being at home" in the world. The expressed worldviews of many people in a community flow together to support collective action and thus do in fact create a world, though not exactly the world anyone envisioned. In part, this unforeseen quality is owing to the size and complexity of society; in part, it reflects the constraints nature imposes on human activities even when nature has been "tamed" by prior human activities such as the building of cities, vehicles, and power plants.

In other words, human representations of the world come to life, but not in a self-fulfilling way. Instead, they undergo strange transformations as they interact with society and with an objective reality that can never be fully articulated and yet forms the unavoidable context of communication. It is quite common that texts promoting one version of reality provide the justification for collective action that works to invalidate the original texts. For example, speeches, discussions, and ads promoting the cornucopian myth that the world's resources are inexhaustible spur rapid consumption that accelerates the arrival of conditions of scarcity and thereby puts the lie to the cornucopian myth. The inadequacy of prior texts becomes apparent, and people generate new texts, though not without conflict and environmental disruption and not as soon as many would wish.

The intransigence of the objective world with regard to our representations necessitates continual revision of those representations through a dialogic, interactive, collaborative process. Meaning is contested by different groups in society, but the relative power of these groups does not in itself *explain* the "career" of any particular meaning because meaning resides in symbols even if they are constituted through the arrangement of sequences of signs. When our symbolizations of the world fail the test of time, we recognize this sooner or later, though at times we are more attentive to the social struggles that surround such failures, and at other times attention is diverted away from the problems as long as possible. Such diversion naturally is driven by those people who are emotionally or financially invested in failed or failing representations (the signs masquerading as symbols that will sooner or later be revealed as signs) because such signs consolidate their defenders' social power even as they generate "externalities" in the form of human suffering. Reality is therefore defined through a social process, but this process is not free floating and arbitrary in its long-term course; it is ultimately tethered to objectivity. The special utility of symbols is that they form the strands of this tether, while signs serve a complementary function of creating a virtual space and time in which many realities can be hypothetically constructed. If the space of signs is inherently counterfactual, the space of symbols is inherently factual. We come, then, to the argument that unites the concept of the extensible self to our framework of communication theory: my plea for breadth in four different facets of communication.

Communicational Breadth

Insofar as communication varies between arbitrary and motivated forms and these forms must be coordinated so as to minimize human suffering and environmental degradation (which is also an indirect cause of human suffering), and insofar as such coordination occurs, as Habermas has shown, through a cycle of communication and action, and such cycles lead too often to human suffering (indicating failures of communication), we can specify four forms of breadth that are constitutive of good communication.

1. The more people included in the "we" who are undertaking the *reflexive* process of representing the world, the broader the range of perspectives and the more productive the debates can be.

2. The more places we discuss in this reflexive process, the more likely it is that we will foresee the future consequences of present actions in those places that concern us the most.

3. The more flexible we are in *how* we represent the world (in terms of our repertoire or lexicon of signs, symbols, and signals), the more likely we will be able to act collectively on our reflexive understanding.

4. The more we overcome a bounded sense of self and recognize the extensibility of the self, the more likely we are to develop nonarbitrary symbolizations of the world in which we live.

This chapter therefore argues for four kinds of breadth in communications: social, geographical, rhetorical, and moral. Note that the term *we* appears in all four definitions, but in the first definition the *contingency* of *we* is specified: "we" are those who are participants in a discursive field, so social breadth (the inclusiveness of the discursive field) is the foundation for defining all other forms of breadth. Habermas's ideal speech situation is reconstituted as a goal of maximal possible connectivity among persons. Rather than focusing on language, we focus on the boundaries of the self. Each other form of breadth in the list implies, in turn, not just a way of delimiting who is communicating, but *how* they are communicating—a particular way of being in the world.

Social breadth involves participating in inclusive communication situations that allow the widest possible range of participants and include them with as little control as possible regarding when and how they may contribute to the discussion (noting that a complete absence of control probably needs to be avoided because it can lead to a bout of pointless name-calling or other breakdowns in communication, as in the Internet religion debate). New technologies reopen all debates with regard to social structure, but we should not enter debates with the presumption that only one type of social structure (or no social structure at all) is good. Social breadth implies the inclusion of people in various social roles while acknowledging that the communication situation reconfigures those very roles.

Geographical breadth involves the appreciation of concerns from as wide a range of places as possible, taking into account the fact that people are increasingly bound up in the affairs of distant places through global flows of goods, information, money, and people. Any kind of political deliberation in particular should include the concerns of distant others who would most likely be affected by our actions. The prevalence of the "spatial fix" has meant not only that problems are "solved" by redirecting the impacts from areas of political power toward areas of weakness, but that people maintain the legitimacy of this redirective practice by dismissing the complaints and concerns of those with comparatively less power (Harvey 1982). Therefore, drawing a geographically broader range of participants into a discussion often if not always produces areas of agreement that are intersubjectively more valid than when the discussion is place bound, provincial, or parochial. A foundation of trust between spatially separated agents must be developed and actively constructed through two-way rather than one-way communications.

Rhetorical breadth emerges when the participant in communication recognizes that there are various ways to weave a whole text (a complex symbol) from the disparate elements of experience, to make sense of things, to make a work of beauty, or to construct a meaningful narrative. This form of breadth resolves apparent philosophical debates by recognizing that you and I may state things quite differently and both be *right*. Symbolic correspondence is nonarbitrary, and therefore disagreements need not be solved through force, on the one hand, or through postmodernist relativism, on the other. By acknowledging that several arguments may capture different elements of the communication context (reflective of different horizons), we can recognize the validity of opposing viewpoints but also try to improve their correspondence with their selected part of the context. Furthermore, rhetorical breadth helps to defuse moralistic judgments that normally arise from the overrun character of symbols noted by Barthes and to shift the ground of debate from morality to practicality.

Moral breadth, in very general terms, is one's sense of commitment to the well-being of the potentially infinite array of others with whom one's lifepath is intertwined along its infinite branching border. To see this point, let us consider a soldier following orders and a pacifist protesting a war. Both are communicating, and they may even share many of the same ideals. Furthermore, both individuals are responding to modern conditions of disembedding and distanciation by developing a firm commitment to a social network that is too large to know or comprehend directly. The crucial difference is that in doing so they prioritize different geographical scales of "community." When one is faced with a choice between the pacifist's global-scale community and the soldier's national-scale community, the best guide to morality is a commitment to symmetry between the scales of selfhood: one's sensory awareness should roughly match one's ability to act. If one eats food picked by hands in Mexico, wears a shirt made by hands in China, and drives a car fueled by petroleum from Saudi Arabia, then one's agency extends to these places, and one's willingness to communicate discursively should also extend as much as possible to these areas, or at the very least to "representatives" of these areas—immigrants, students, and travelers who bear something of the communication context of these places (recalling that context is defined by connections, not by locations). These "representatives," most important, can share with us their view of the impacts of our actions. By engaging in conflict without dialogue, the soldier is at best counteracting one force of oppression, but is unfortunately always an agent of oppression in his or her own right. A just world cannot be established through "defensive" force, but only through engaged discussion. By recognizing my long-distance communications as parts of myself, I can begin to assume responsibility for the range of places I affect through my daily actions, and I can choose the most constructive and least destructive ways of being, doing, and communicating. This fourth form of breadth—moral breadth—is at heart the inclusion of others in the field of care drawn around the self. It is opposed to the narcissism of national isolationism and to the arrogance of imperialism.

What do these principles of communicational breadth tell us about communication? Do they sound an optimistic or a pessimistic note? In particular, they ask, "Can I make the world a better place through my communications?" A postmodern sensibility makes it seem naïve to answer in the affirmative, to argue that communication leads somewhere *better* and that it can be the basis of progress. But assuming I am not willing to discard completely the goal of making the world a better place, I must take this optimistic view of communication. I *must* trust at least some of the communicational contexts at hand to be capable of carrying socially constructive (broad) messages. Even if the media are commercial, the need to communicate rules out over the wish to avoid communication contexts colonized by economic relations. This does not mean I would wish the communication media to be perpetuated in precisely their current form, but I am willing to inhabit them as an extensible individual and extend my agency through them to work with and for others. The question then becomes not *if* I can build a better world through communication and for communication, but *how* I can do so. If I understand the practical implications of communication content and context, I can prescribe better situations for communication, better social futures, and more environmentally responsible ways of living on the planet. Again, what is required of me is attention to breadth in all of its various forms. Extensibility implies responsibility.

Despite the impossibility of knowing the entirety of the space inhabited by the self, my call for breadth implies the very important potential (unrealized unless we make use of it) to bring the space of agency and the space of awareness into rough congruity, to produce a kind of symmetry between knowledge and action. This is definitely not a matter of colonizing the world through discourse, but rather a task of bringing communication to a space already colonized in several ways. This colonized space is not simply geographical space, but the virtual spaces in which a distanciated society is regulated, largely consisting of what Habermas calls the "system." What is implied, for example, is symmetry between the spaces one affects through consumption and the spaces one occupies through online discussion, as well as symmetry between the respect one *demands from* and the respect one *affords to* others. *The purpose of communicational breadth is, then, to assume conscious, reflexive control over extensions of self rather than to allow them to exist like appendages of the system itself.*

Gunnar Olsson argues that "in shared activities like seeing/pointing/speaking/hearing/reading/writing, we become extensions of one another; no body is a self-sufficient entity onto [*sic*] itself, but always a double in need of the other" (1988, 126). Extensibility is not just a part of being human; it is a part of each human being. Extensibility is the geographical aspect of personal identity. I am my extensible self because when I say "I," the self, already constituted through extensibility within a distanciated society, is automatically invoked as something *meaningful.* So my prescription for communication situations actualizing symmetry between action and knowledge gathering is also a prescription for fulfillment of the self.

If this goal is not to be perverted, the main challenge is obtaining rhetorical breadth. To see this challenge, we must track current geographical discourses and reveal their adequacy and inadequacy with regard to the communication issues I have sketched here. Let us start with my own rhetoric. I have placed the globalization of communication in a somewhat tragic light: the economic, political, and military systems that create for us a certain level of predictability and security also incur responsibilities, although we are generally remiss in meeting them. The more we transcend the constraints of local conditions, the more our duty to engage in dialog with others expands. It is a classic tragedy; our very nature is to look for ways to ignore or deny our far-flung responsibilities, yet doing so has grave consequences for others and ultimately for ourselves. But this discussion also suggests the motif of cosmopolitanism, a concept that is inherently romantic—the idea Marshall McLuhan captured in the term *global village* (1962, 31). It would be naïve to hope to transcend the chasms in culture and language between the world's people if transcending them were not, in fact, necessary. Because it must be done, we are faced with the question of how best to communicate if we wish to promote a constructive and reconciliatory interaction between people who are very different but who are of necessity involved in each others' lives. According to my description, the moral situation of the extensible self includes both tragic and romantic elements. (This is not an observation, which is to say a statement of presumed fact drawing on and contributing to the objective sphere, but rather a prescription for how best to understand communication that must be ongoing and must overcome differences.) In other words, it is my intersubjective judgment that we should see ourselves as faced with a situation that is *both* tragic and romantic. This mix of viewpoints suggests irony, a third narrative appeal. Let us obtain a better sense of what *tragic, romantic,* and *ironic* may imply in terms of action and communication by drawing on rhetorical theory.

Narrative Modes

Let us recall the point of chapter 3 that statements are motivated by aspects of experience. Without this motivated quality, all statements would be as arbitrary as the words that make them up. We would not be able to choose between "it is raining" and "it is sunny," but would have to settle disputes over the weather (and everything else) simply through force. Instead, statements, texts, and discourses are symbols, and although force plays a part in communication, it would be a gross oversimplification to reduce communication to social conflict. We can resolve debates by obtaining knowledge about the world (which is a good thing for those who take an interest in learning).

Many social scientists have mistakenly followed the lead of semiotic theory in studying culture as a collection of sign systems. If every string of signs were itself a sign, we would be mystified by any sentence we had not previously encountered. There would be a need for a massive dictionary that contained every possible statement (and of course another dictionary defining every definition in the previous dictionary, and so on). In addition, we would have no way of translating a phrase from one language to another unless it was already translated or could be translated word for word. Most disturbingly, we would be faced with two unpleasant alternatives to disagreement: radical relativism (which prohibits coordination of action) and force (which coordinates actions despite disagreement).

If an English-speaking Canadian looks out the window and says, "It is raining," we suspect that his Francophone colleague will assess the weather by saying, "Il pleut." The fact that we can use perceptions framed in one language to predict observations in another language indicates a substrate of shared experience that transcends the different frameworks imposed on raw experience by language. Less trivially, we can apply this idea to infer that a book by Karl Marx translated into Spanish or Hindi can theoretically convey essentially the same symbolic representation as the English version and that the global debate on climate change can potentially lead to international regulations that benefit everyone.[1]

What exactly is carried forward through many translations of a particular text, whether it is about today's rain or the long-term trend of global warming or any other subject? Each text has a specific overall meaning made up of the sum total of its words or other elements, but this (symbolic) meaning is impossible to specify precisely without actually repeating the text. Yet other symbolic elements of every text are more general and are shared with a great number of texts. These elements are tropes (such as metaphor and metonymy) and narrative modes of appeal: tragic, comic, romantic, and ironic. Even more specific are ideologies, such as Marxism, positivism, and humanism. Here I concentrate on narrative modes in order to help us better understand

---

1. My position is opposed to the postmodern position adopted by Marvin Waterstone, who condemns geographers for trying to sound an alarm about the risks of global warming: "It is only by adopting a particular construction of 'global' warming, one which views nations (and subgroups within those nations) as being 'all in this together,' that Harman et al. can begin to postulate something like *public risk* or *the public good,* and suggest that scientists are in a privileged position to identify and evaluate these matters" (1998, 299).

the nature of some well-known worldviews in geography. This section shows rhetorical breadth as a necessary adjunct to geographical and social breadth and ultimately as a basis for moral breadth. In regard to existing geographical writings, this means overcoming the stigma that has been placed on romantic and comic appeals, without abandoning the tragic and ironic voices that currently dominate the scholarly discursive field.

Rhetorical theory includes particular tropes or turns of speech. We may more accurately think of them as "turns of thought" because they arise in communications that are quite remote from speech, such as visual art and even landscapes. In this connection, James Duncan (1990) has shown the use of synecdoche in the Kandyan landscape, and David Lowenthal (1985) has exhaustively explored the metaphorical links between history and places and the past. Here I focus on the other dimension of rhetorical theory, narrative appeals.

The study of rhetoric has revealed four broad narrative frameworks in which novelists, playwrights, historians, and scientists frame their accounts of the world and build instructive tales. Drawing on Northrop Frye (1957), Jonathan Smith (1996) defines these frameworks as *fictional modes*. I prefer the term *narrative modes* because it avoids the implication that these organizing symbols are restricted to imaginary events and settings. Every attempt to express and explain the world depends on one of these modes or, more often, on a combination of modes. The narrative modes are romantic, comic, tragic, and ironic. They are ways of representing the world captured in the concatenation of long strings (texts) of words (signs) to form symbols (selective extractions and abstractions of objective reality). They are thus real aspects of the world, and although each is incomplete, together these modes capture the most general aspects of human existence in the world.

Some novelists and dramatists tell tragedies, some tell comedies, and others weave romantic or ironic tales. These four modes appear also in academic writing, where they are carefully disciplined by the contingencies of methodology and epistemology. The purpose of the narrative appeals is to provide a sense of completion or closure to narratives of all sorts. Indeed, their truth claims (in the Habermasian sense) are based not on objectively verifiable facts (although people invariably marshal facts to support arguments framed in one or another narrative mode), but rather on subjective feelings about this or that situation or event in the world, and intersubjective beliefs about how the world *should* be represented.

Each narrative appeal is *motivated*, like an outline of reality drawn from a different angle, a point easiest to see if we note that each appeal is *successful* in its own way as a guide to action—if not for everyone, then at least for people with certain temperaments. People often seek out narratives that signal their favored action orientation: cautious people prefer tragic narratives, cooperative people seek out comedies, creative people are inspired by romantic tales, and the most observant or perhaps voyeuristic souls opt for irony. This is as true of scientists as it is of novelists or their audiences. For a certain discursive community, then, a particular rhetorical appeal is most likely to be accepted as valid. The theory of communicative action suggests that we separate the subjective preference of narrative appeals from the objective and intersubjective elements of reality represented by many differently formulated accounts. We usually brand narratives of a type we do not favor with terms such as *naïve, cynical, superficial,* or *romantic,* judging and condemning content on the basis of rhetorical preference, as Jonathan M. Smith (1996) has so perceptively argued. One goal in considering the narrative appeals is to recognize that each one symbolizes a valid model of some aspect of experience; none is arbitrary, but in its own way and in particular contexts among certain audiences each serves as a morally and pragmatically valid guide to action. Rhetorical breadth means the open acknowledgment of the validity of appeals one does not favor and the attempt to respond discursively to others in terms of intersubjective and subjective concerns central to the mode(s) *they* are employing. In no way is this call for rhetorical breadth meant to discourage disagreement or debate. Disagreements about the objective conditions of the world must, however, be disentangled from disagreements about the most appropriate narrative appeals to employ in one situation or another. What is not particularly productive is to critique a tragic vision of the world as if it were a failed comedy, to deconstruct a romantic vision as if it were supposed to be a tragedy, and so on. We may want to disagree entirely with the use of a particular appeal in a particular situation—for example, the comic appeal when writing about global climate change—but rational discourse should involve direct *pragmatic* (intersubjective, action-oriented) discussion of why a particular appeal is appropriate or inappropriate in a particular discursive field.[2] The issue should be framed in terms of the interpretive opportunities provided to audiences by a particular narrative appeal, and the actions that can reasonably be expected to arise from such opportunities (although various audience subjectivities will always give rise to a range of interpretations and actions). Rhetorical breadth implies meeting the audience and other "speakers" halfway.

The Appeals

Tragedies are stories that hinge on necessity. Tragic plays and novels generally end badly, but tragic narratives in science are recognized by their emphasis on *boundaries, limits,* or *constraints.* The hero of a tragic novel is often endowed with exceptional strength, intelligence, beauty, or endurance, but that does not exempt him or her from universal limitations that lead to misfortune when fate intrudes with startling suddenness. In the tragic narratives of social science, we find "the actor," "the individual," or "the agent" in place of the hero. He or she is subject to various forces or powers that shape, produce, or overcome personal goals and objectives. Marxism, for example, explains history in a way that tragically excludes or limits the scope of free will.[3] Environmental determinism and Freudian psychology in different ways also tell tragic scientific narratives. What binds tragic narratives together is the com-

2. On the topic of discursive fields, see Duncan 1990, 4–5.
3. In *The Eighteenth Brumaire of Louis Bonaparte,* Marx wrote: "Men make their own history, but not of their own free will; not under circumstances they themselves have chosen but under the given and inherited circumstances with which they are directly confronted" ([1852] 1973, 146). Although this quote has supported many tragic extrapolations of Marxist theory, it is worth recalling that Marx generally wrote in the romantic rather than in the tragic mode, and the mode shifted when Marx's followers became skeptical about the possibility of radical social change.

mon thread of cautionary advice. The overwhelming tone is sober recognition of constraints, yet there is a desire here as well. Tragedies symbolize the desire to see the world coldly and clearly, without illusions bred of complacency and without the warming fires of hope. Yet that "clarity" is always, like all other appeals, a symbolic construction. Hope is real (people feel it), and, perhaps more important, it is useful (people who feel it are inclined to act more effectively on their own behalf). We must interrogate tragedy with regard to what it does (particularly if it discourages effective action) rather than simply accepting it at face value or assuming it is politically superior to other perspectives.

Comedies delight in the resolution of contradictory forces. Contrary to the popular use of the term, comedy does not need to be funny. What it does above all is show that the obstacles to satisfying individual interests can be held at bay, circumvented, or co-opted. The happy ending is thus integral to comedy, and comedy symbolizes above all the desire for resolution and mutual satisfaction. A comic fictional tale ends in the discovery that evil is less potent than it at first appeared, and adversaries are overcome by cunning or by their own greed and stupidity, but generally they learn their lesson rather than being annihilated. In academic writing, this appeal can be identified mainly by the message that conflicting interests and objectives can be reconciled and harmonized, as in neoclassic economic theory and various applications of ecological theory, at least in ecosystems without human involvement.

Romances are sagas of transcendence. Romantic fiction of course includes stories of men and women falling in love despite intransigent parents and the like, but romance's essence reaches beyond gender dynamics to all situations in which the hero rises above obstacles to achieve his or her desires. Obstacles are not simply pushed into the background (as in comedy); they are annihilated or banished, consigned to another period of history or to another place. Romance's promise of a better future is a powerful foundation for social change and thus has found its way into several branches of social science, though often mixed with tragedy (as in Marxism). Romance symbolizes the fact that limitations may change or be changed, and hence it captures a facet of reality hidden to tragedy. Romance also, at the same time, symbolizes a desire for something better. In this regard, it is a less-satisfied mode than comedy. Romance is found in academic writing as futuristic speculation, counterfactual supposition, and even utopian visions. Transcendence of social limits is, naturally, rare, and transcendence of natural laws is illusory, so this appeal may deceive or mislead.

Finally we encounter *irony* or farce. Ironic narratives encourage a sense of distance or detachment. Irony may be humorous or cynical, but either way it differs from comedy and tragedy in that it takes a markedly detached view and encourages its audience to do the same. It is less emotional than the other appeals, neither cold like tragedy nor warm like comedy and romance. It arises from a subjective desire to be "above it all," to achieve what Thomas Nagel (1986) calls a "view from nowhere." In this regard, irony's appearance of logic is deceptive because irony is often self-indulgent. Irony views existing struggles or conflicts from an apparently impartial viewpoint because it offers more than one viewpoint. This book, in communicating about communication, is itself ironic. The strength of irony is that it steps back from existing conflicts, substantive and rhetorical, but this strength is also its greatest weakness because it easily lapses into sophistry. Struggles between humans and their environments and between various human groups become spectacle, and the author and audience indulge in voyeurism. This voyeurism can be constructive in that it permits detachment from narrowly defined interests (e.g., John Rawls's "veil of ignorance" [(1971) 1999]), but it can also inhibit a sense of involvement that respects the grounding provided by objective conditions such as asymmetrical economic flows and how such grounding may be transcended through actions rather than merely through symbols.

To think of rhetoric analytically is to reflect on basic modes of understanding the world and hence on basic ways of *relating to* the world. Because we cannot fully control the world, it often happens that our "way of relating to the world" appears imposed on us by the world itself; the world seems this way or that, despite our will or desire. It seems as if we apply rhetoric retrospectively to understand how the world "is," but we in fact intertwine events into a kind of formulaic story, a skeletal narrative that allows our experiences to be understood, not recognizing that more than one such narrative can be applied in most situations. Hayden White (1996) recognizes this kind of relativism (based not in a dismissal of objectivity, but rather in a recognition of objectivity's plurality) in regard to the writing of history, and the conclusions he draws about historiography illuminate how geographical narratives also serve to ground geographical phenomena in lived experience and the objective materiality of the world while closing down recognition of other frames of thought.

Narrative and Historiography

Until the early twentieth century, a history meant a story with a beginning, a middle, and an end. The story usually had protagonists and some kind of climactic event such as a war or a change of political regime. Histories concentrated on individual people, actors who determined the course of world events. The temporal, sequential character of history itself seemed to dictate this kind of exposition. With the rise of social science around the turn of the twentieth century, historians began to question this narrative approach and to call for a more scientific, analytical approach. Beginning with Marx's *Eighteenth Brumaire of Louis Bonaparte* ([1853] 1973), historians often adopted structuralism in the form of historical materialism as an increasingly popular alternative to biographically oriented histories. There was an associated decline in professional respect for the narrative approach. Most recently, however, a number of historians have asserted the need to appeal to lay persons, which leads back to narrative, but more self-consciously.

Hayden White supports history's "return to narrative" but takes issue with the reasons motivating most of the supporters of this movement, which apparently have to do with the desire to reach a broader audience by making history more palatable. He writes, "The suggestion is that [narrative] is merely a matter of dressing up their findings, produced by the application of social scientific methods to their objects of interest, in the garb of a story, in order to make the results of their research more palatable to their lay audience" (1996, 59). In opposition to this view of narrative as a matter of packaging that can be applied to realities revealed by social science, White cites four decades of research showing that "narrative is an expression in discourse of a distinct mode of experiencing and thinking about the world, its structures, and its processes" (1996, 59). Rather than an arbitrary form of ornamentation dictated by taste or fashion, narrative is a symbol of the real assembled of the arbitrary *signs* called words; through the multiplication and arrangement of signs history ceases to be arbitrary and instead corresponds to aspects of the real.

Before pursuing his argument, White presents two critics of the use of narrative in historiography. Fernand Braudel interprets narrative as a "philosophy of history," which conflates a period in time with the interactions of a few extraordinary individuals whose lives contain conflicts and climaxes of a kind found only in the lives of fictional or mythical heroes. Roland Barthes claims similarly that "historical discourse is in its essence a form of ideological elaboration," a mythology (quoted in White 1996, 60). Both Braudel and Barthes reject narrative as a legitimate device for telling histories on the grounds that narrative makes the world into a kind of theater or spectacle. History appears to unfold with key players and a plot, as if it were authored, yet this appearance is imposed on history by an author who therefore promotes a fundamental misconception about the way things happen. What historical narratives produce, in their view, is *historical realism,* akin to realism in fiction, which frames events in a master narrative that leads up to a lesson and in the process generally promotes bourgeois interests by minimizing the significance of ordinary people and their interests. On this account, progressive history must subvert the narrative's authority in order to favor instead more structural and analytical forms of awareness that do not imply the same kind of actors, author, and authority. Nonrealist historiography would ostensibly draw closer to the reality in which people actually live their lives. Rejecting narrative, it would prepare people for actions that would shape history in progressive ways.

White admits that narrative structures must be purged from historiography if we, like Braudel and Barthes, assume that "the kind of coherence imposed by the imposition of a plot structure—of the sort met with in mythical and fictional discourse—on a given set of historical events has no counterpart in reality" (1996, 65). But he questions whether narratives are necessarily imposed and arbitrary structures. He suggests instead that they "work" because they are shared forms of awareness. Plots are real in that what has been understood as a sign (the narrative) is in fact a kind of symbol, with the symbol's nonarbitrary (albeit limited) link to various realities. Interestingly, it is Georg Lukács who provides White with his key question: "Is it possible . . . that not only 'climaxes,' but also the kinds of *plots* that utilize the 'climax' in different ways, *occur* in 'reality' or at least in 'historical reality,' as well as in 'fiction' and 'myth'?" (White 1996, 66, emphasis in original). The plots of which Lukács speaks include the epic, romance, tragedy, comedy, and farce, a list strongly reminiscent of the narrative appeals mentioned earlier.

White's argument turns on a conception of reality that is clearly not empiricist. Like Habermas, White assumes that reality includes the metaphysics of observation and of sharing. His source of inspiration at this juncture is David Carr, whom he paraphrases as arguing that people inhabit a "sociocultural world that is structured narrativistically and intend their actions in such a way as effectively to make of them the kinds of actions about which 'true' stories can be told" (1996, 67). Thus, narratives can be real not because they match some putative objective world to which an expert or social critic has privileged access, but because they symbolize life in a world inhabited by creatures that *must* know themselves and others through stories. This is a refinement of the idea of reflexivity—it explains how reflection is organized. These stories that people use to reflect on the world and on human actions in it are always synthetic in the sense that they draw together subjective to objective and intersubjective experience. Rejecting narrative in history is tantamount to narrowing the ground of "relevant" experience to the realm of the intersubjective, a move similar to that taken by the scientific historians who tried to exclude subjective and intersubjective dimensions and to reduce history to objective fact.

Even more important is the link between historical narrative and future action. Humans have agency, which implies that they not only act, but act purposively—in other words, they think ahead. This thinking ahead implies "emplotment"—a strategy for weaving stories out of experience, making narrative *symbols*—that is not only part of the historian's retrospective vision but also part of the projective vision of people living their lives in real places: "human agents *prefigure* their actions as narrative trajectories, such that the outcome of a given action is at least *intended* to be linked to its inauguration in the way that the ending of a story is linked to its beginning" (White 1996, 67). *Therefore, history's ideal form is the retrieval of the lost prefiguring emplotments of agents who are no longer present.* It tries to do this by examining the way people acted and whatever bits of their communications have been preserved, then building a story with the elements of narrative that make the most sense. Historians therefore are not (only) inventing a story when they engage the past through "emplotment"; they are *discovering* a story that people intended to live (or thought they were living). Historians also benefit from hindsight because they can see how some parts of that story look in the light of subsequent events and how people have reflected on the situation. Still, White is not entirely satisfied with this explanation because it leaves no basis for distinguishing history from fiction, myth, and fable.

Insofar as various plot types are at historians' disposal, the act of writing a *narrative history* still implies a choice between the forms of narrative, and this choice carries ideological content. So how does a historian choose between tragedy, romance, farce, epic, or comedy, and what makes that choice particularly appropriate to writing history (as opposed to fiction)? Furthermore, how can we reconcile two or more narrative accounts that "appear to be the same set or sequence of historical events, when the stories told about them are manifestly different, contradictory, or even mutually exclusive?" (White 1996, 69). White provides examples of authors whose ideological stance led them to prefer one kind of story: Hegel's preference of the comic over the tragic, Lukács's preference of the epic over romance or comedy, Marx's preference (in the *Eighteenth Brumaire of Louis Bonaparte*) for farce over epic. Such choices may be viewed in terms of narrative preference rather than distortions of fact, and rhetorical preference is not necessarily arbitrary. Valid reasons based in recorded information about what was

form of expression:	form of content:
grammar syntax lexicon	"the facts" names measurements locations spatial connections sequence of events
substance of expression:	substance of content:
farce romance tragedy comedy epic	class struggle heroism

5.1. Interpretation by Hayden White (1996) of Louis Hjelmslev's taxonomy of discourse.

said and done in the past might lead a historian to choose one "figuration" over another. Such reasons, White argues, point to the difference between history and other modes of writing, such as fiction, myth, and fable.

To discover what it is that distinguishes history from other modes of figuration, White adopts a model by Danish linguist Louis Hjelmslev. This model's analysis of discourse employs a twofold dichotomy—form versus substance, on the one hand, and expression versus content, on the other. Hjelmslev's framework, as interpreted by White (figure 5.1), permits history's unique qualities to be isolated even while it recognizes that history tells stories using the same rhetorical devices as fiction and other types of writing. Furthermore, it indicates how different histories can use the same facts but arrive at different interpretations of the meaning of those facts. What we conveniently refer to as the "meaning" of a particular historical account is, in this framework, a composite of the substance of expression and the substance of content (figure 5.1, bottom row). The author must accept as given the two types of "form"—the rules of his or her language (spelling, grammar, etc.) and the recorded "form of content": names, measurements, locations, spatial connections, and the sequence of events. He or she must choose some "substance" (narrative appeal) in which to communicate, according to his or her understanding of events. This nuanced view of the historiographic project permits us to recognize that narrative histories can be both *constructed* (through choices of signs and symbols, the latter including the "substance of expression," or narrative appeal) and *true* with regard to known facts and shared experiences. A disciplined approach to form is the mark of history, and this approach leaves open to debate the differences in perception resulting in the range of substance.

White concludes that Marx wrote history when he wrote the *Eighteenth Brumaire of Louis Bonaparte,* although (and perhaps because) he invested his story with the figurative meaning (substance of expression) of farce. It was the notion of class struggle that motivated Marx to interpret the facts in the substance of a farce, and class struggle is widely recognized "by all historians of every conceivable kind of political persuasion or ideological orientation" (White 1996, 78). Another historian might not choose to discuss class struggle, but class struggle is a recognized pattern historians use to explain how events occur. Of course, this justification makes any *new* explanation suspect until proven "historical"; then the key to legitimacy is not in the work itself, but in the community of historians. Grounding the writing of history in a discursive community serves the purpose of broadening individual perspectives and yet avoiding a fixed, dogmatic approach. In the case of the *Eighteenth Brumaire,* the substance of content, class conflict, can be seen as valid in terms of Habermasian intersubjective validity claims judged by historians in terms of a particular genre of history. Its application to this particular situation and the extent to which it accounted for any or all of the related events can still be disputed. Others writing about the same events in France (specifically Victor Hugo and Pierre Joseph Proudhon) were, in contrast, writing fiction, White argues, because they based their interpretation of events (substance of content) on the idea that Louis Bonaparte was an exceptional person, and such reliance on the "great man" explanation is not part of the repertoire of currently acceptable historiographic content.

Now it is questionable whether selecting one's substance of expression on the basis of class struggle automatically entitles one to qualify as a writer of history (after all, some who do the same thing are trained as geographers and write structuralist geography, not history), but Hjelmslev's framework serves as a means of recognizing that each of the four narrative appeals is theoretically independent from the others. Two writers may agree on the substance of content (say, class struggle) but arrive at different narrative modes (substances of expression) such as tragedy and romance. This is as true of geographers as of historians.

Of course, all this does not resolve any particular historiographic debate. You and I may agree on the historically formative role of gender relations, but whereas you might emphasize the tragic nature of these relations, I might emphasize the comic nature of the same relations. Our ultimate goal may be the same (to motivate other persons to act against gender-based injustices), but differing subjective response to narrative appeals may cause us to craft radically different narratives in support of the same pragmatic objective. This happens because we try to understand our audiences by looking "within" at our own responses to narrative appeals, yet not everyone responds the same way to the appeals (or to any other aspect of communication). Narrative appeals as symbols of the world are particularly complex, like religious beliefs; therefore, differences in the writing of history are produced by more deep-seated differences in personality. White does not assert that these differences can be resolved. He merely asserts that we can write different, equally valid stories, all of which are histories related in a nonarbitrary way to the real, and that this multiplicity does not mean "anything goes."

Rhetoric and Academic Debates

In fields of study less well known for their narratives, debates often emerge out of the different rhetorical inclinations I have just outlined. The coexistence of different narrative appeals is perhaps the most important factor inhibiting mutual comprehension in the social sciences and humanities.

As Jonathan Smith points out, geographers who advocate one the four narrative appeals frame their critiques in a predictable way, based precisely on their narrative preference (1996, 10). A geographer who prefers the tragic mode will condemn texts written in a comic mode as complacent, while judging texts in the romantic mode as idealistic and works in the ironic mode as dangerously deluded. A reader who prefers the comic mode will tend to see works written in the romantic mode as reckless, in the tragic mode as pessimistic, and in the ironic mode as alienated. Those who prefer the romantic mode will often feel concerned that other approaches are too conservative (whether tragic or comic works) or else too cynical (ironic works). Finally, a geographer attuned most strongly to irony will often dismiss other approaches as self-aggrandizing, pretentious, or naïve (when reading romantic, tragic, and comic appeals, respectively). Expecting these general responses assists us in understanding the responses any particular work will engender within the scholarly community of geographers, and it also helps us understand why geography is not better appreciated as a discipline. Internal epistemological, methodological, and ontological fractures remain unbridged because of a communication failure, thereby preventing outsiders from getting a clear idea of what it is that geography is about, so their interest is often frustrated.

Academic writings are motivated, nonarbitrary representations of the world, not only in regard to the "facts" they marshal (the form of their content), but also in regard to the interpretive frameworks they employ (although, as we have seen, it is not possible to resolve disputes about these interpretive frameworks). That makes them symbols, and we already know that symbols are not only partial, but also overrun (treated as quasi-sacred and freighted with subjective desires). They are therefore susceptible to being judged as morally bad and "out of place." This suggests that most of the time when people argue about the world, they are in fact arguing about different symbols of the world. In doing so, they are unaware that many symbols of a single thing may all be "right," which is to say they all correspond with some aspect of reality in a nonarbitrary fashion. They either reject a certain appeal as false and deceptive or they deny the idea that any description of the world can be true.

The virtue of what I have called "rhetorical breadth" is that it allows discussions to move beyond sterile debates based in narrative preferences and potentially to deal with more important questions. This progression in turn facilitates moral breadth and ultimately a more just and equitable society. At the same time, it also helps people appreciate worldviews arising out of different cultures (which facilitates geographical breadth) and different subcultures (which facilitates social breadth). These extensive repercussions reflect the grounded nature of communications; each of the four forms of breadth corresponds to a particular broadening of the juncture between lived experience in places (physical and virtual) and the act of sharing those experiences through signs, symbols, and signals, an act that in itself constitutes other places (physical and virtual).

Geographical Views of Communication

At this juncture, I analyze some worldviews from the geographic literature to see how they position authors and readers relative to the four narrative modes. I have chosen geographical writings that are responsive to the expanding influence of communication in daily life and have drawn from the writings of geographers because that is my area of expertise and because I believe strongly that geographical scholarship matters. I also want to demonstrate how geographical writings express the erroneous idea that texts and discourses can be understood solely in terms of how they strategically and politically constitute various worldviews. On this account, the meaning of communication can be understood only in terms of social power relations, and the geographical dimension of meaning is always dependent on social power relations insofar as differently empowered groups are positioned differently in space. Yet, in this view, the grounding provided by the richness of place—as a mix of symbolic, social, and physical elements—is missing.[4]

This *ungrounded* view of communication is central to a tragic or ironic worldview, or both, that inhibits the emergence of moral breadth.

4. For detailed discussions of place as an experiential and cultural phenomenon, see Entrikin 1991, Relph 1976, Sack 1997, and Tuan 1977.

The geographers I have chosen for this discussion—Peter Jackson, Gillian Rose, Tim Cresswell, Gerard Toal, Don Mitchell, and David Harvey—are generally excellent writers and researchers, but their ungrounded notion of place is illustrative of a general pattern in geography with regard to communication. Their bias does not qualify as a "paradigm," but it is quite prevalent and in fact serves now as a standard of acceptable scholarship. Discussing the "compromised," "political," and "contaminated" nature of discursive constructions helps scholars to be acknowledged by other scholars as doing important and high-minded work, and the usefulness of such work is taken for granted as long as social power relations are "excavated" from discourses, and texts are thus "destabilized" or "problematized."

My concern with this ungrounded geographical writing is pragmatic. Geographic writing can and should be more effective as a tool for reflexive thought and as a means of intervening in social processes. I concur with Gunnar Olsson that "The ultimate test of our understanding is . . . less in explaining the past than in shaping the future" (1975, 6). We must be aware that the tragic-ironic view of communication initiated sixty years ago by Horkheimer and Adorno and reiterated again and again has failed to produce significant political or social change. Any number of newly coined terms cannot disguise that fact. I am not questioning the stated objective of "intervening" in society and achieving what Habermas referred to as "emancipatory" objectives, but the rhetorical scope of this critical project needs to be rethought if it is to live up to its goals and objectives. Geography serves as a case study for other realms of discourse, both academic and nonacademic. It indicates that *rhetorical biases can operate after the fashion of paradigms,* silencing nonconforming texts and structuring who can and cannot join in discourses (Kuhn 1996).

We start with Peter Jackson, whose influential *Maps of Meaning* (1989) developed the argument at the end of the 1980s that communication is political, implicating language in intergroup struggles of various types at various scales. Jackson's "politics of language" draws on Giddens's theory of structuration, but in a limited way, highlighting mainly Giddens's ideas regarding social norms and sanctions, both of which emphasize social power and political struggle (Jackson 1989, 157). Thus, Jackson writes: "The structuring of language into systems of dominance and subordination, as described by Giddens' theory of structuration for example, provides a way of understanding how the negotiation of meaning between groups becomes sedimented into more permanent structures and relations of inequality" (1989, 161). As an example, he discusses how states and scholars designate certain variants of a language as "standard" and other variants as "nonstandard." This kind of hierarchical prioritization of dialects obscures the continuous variation of language through space, making some people "right" and some people "wrong" the moment they open their mouths. As we know, signs themselves are arbitrary; what is really at issue is the symbolism of social *membership* carried by particular variants of the acoustic sign. This observation is intriguing because it reveals the way subtle variations in a sign system become symbolic of social hierarchy; when I say "po-tae-to" and you say "po-tah-to," each of us is *symbolically* locating himself or herself in a social space—our objective meanings are identical, but our subjective claims regarding group affiliation differ.[5]

Jackson's attention to power and language reveals an important aspect of the link between lived experience and representations of the world. He explains that when authors describe different social worlds, such as the rural and the urban, the rich and the poor, they encode what they write in such a way that it is accessible to a particular group and that it reinforces that group's evaluation of the social and physical landscape. At the same time, this encoded reality denies other groups' views. A Jane Austen novel serves as an example: "The novel conveys its moral purpose as much by language as by plot, by the juxtaposition of words and phrases that convey the opposition between conflicting social worlds: manners and morals, personality and principle, wit and wisdom" (Jackson 1989, 162). The novel legitimates one model of the world as well as the group that benefits from the adoption of that model.

To round out his engagement with communication and social

5. For more on this phenomenon, which demonstrates that verbal signs are also symbols of identity, see Chambers 1995, Labov 1972, and Saville-Troike 1989.

power relations, Jackson explores the use of language by geographical explorers, specifically the act of naming features of the landscape. Naming can be a means of claiming a landscape for a group even before the group sets foot on the land. This symbolic conquest is tied, of course, to a refusal to recognize prior occupants such as indigenous people and the names they have previously attached to the place. Imperial history and geography are therefore means of "establishing proprietorial claims through linguistic association with the colonizing power" (Jackson 1989, 168; for similar arguments, see Herman 1999, Lewis and Wigen 1997, and Olsson 1992).

Jackson shows us clearly that social power is expressed in terms of the names people choose for places, what dialect they consider "good" or "right," and how they assemble various texts. Despite the validity and value of these observations, Jackson's framework poses some problems. As a body of argument, it situates communication in a two-dimensional space, a space that always has at one pole the poor, the popular, the vulgar, the unsanctioned, and the colonized, and at the other pole the wealthy, the respectable, the polite, the legally sanctioned, and the colonizers. Real societies are not this tidy. Some immigrants are wealthy, some colonizers are vulgar, some respectable people are poor, and so on. How, what, and why do these ambiguously situated people communicate? What do their communications *do*?

Further difficulties arise from Jackson's view of language as a kind of weapon deployed in class struggle, ethnic conflict, and geographical conquest. Through his writings, we perceive a world where communication maintains and solidifies the power of the strong over the weak in varied ways. This is meant to be a clear window onto communication— a higher truth than communication's "face value"—but it is (like all symbols of the real) only a filtered or partial view. In particular, the author leaves no place in his theory for communication to bring people together—geographically or socially. The comic idea of communication as a means of social integration and reconciliation may seem naïve, but it is in fact *necessary*. To write a book like Jackson's is to try to forge links between the powerful and the powerless. The author certainly has power, and the people he wants to help are those without power, yet if we believe the author's portrayal of communication, his own book simply should not be possible.

Jackson makes the comic assumption that mutual understanding can be reached (because he presumably wants people to agree with him) and the romantic assumption that oppressive communication can be overcome (or else why bother to expose communication problems?), but he does not speak of these assumptions or provide a hint of how they fit in with the tragic-ironic view he communicates so persuasively. In other words, he has failed to include his own actions (as a communicator) in his map of communications and thereby has demonstrated the need for a broader conception of rhetoric. His own communicative actions lie outside the scope of his theory because the narrow horizons of a thoroughly political model of communication are too confining. Communication must be understood not as a form or means of struggle, but also as a means of communal world-building and just activity, with all of the ontological and rhetorical complexity that implies.

Many other geographers have followed Jackson's lead. They, too, want to communicate in a socially constructive way, yet their "critical," "nuanced," and "decentered" evaluations of communication draw a bleak picture. If communication is always destructive rather than constructive, then the act of pointing out this situation (being further communication) has no logical or moral justification. Tragic and ironic views of communication bracket out the speaker's stump and obscure the possibility (indeed, the inevitability) of different social worlds, parallel to the world of any one reader. They start with inequality and injustice in physical space, discern a kind of template in virtual space, then oddly forget that this virtual space includes their own engagement with communication, social relations, and reality—their intent to *reach* others.

What specifically is entailed by this critical approach I am critiquing? Agency may be relocated to the superorganic level, to abstractions such as capital and the state, a Marxist tragedy that denies personal agency. More common now is ascribing to people the ability to resist or negotiate with power. Jackson and many other "new cultural geographers" focus on communication as an instrument of domination and oppression. They do not entirely deny individual power or agency, but they either minimize it or, increasingly often, reduce it to the motif of symbolic struggle. They identify multiple axes of power, multiple forms of resistance, and multiple battlegrounds, but the increasing refinement of this communication-as-conflict model leaves intact the central premise that communication is an instrument for separating, dividing, conquering, and submitting one group to the will of another. This tragic-ironic worldview is blind to the ways that communication neither succumbs nor resists, instead operating in a completely different space, according to different rules than the binary dynamic of the struggle for power.

Edward Relph asserted that the aim of humanistic geography was "to illuminate the places and environments of the world, and to interpret them, good or bad, so that others can appreciate their significance" (Relph 1989, quoted in Rose 1993, 50). Relph's comic perspective has predictably come under attack from geographers. Feminist geographer Gillian Rose, for example, rejects Relph's assumption that a geographer can illuminate the world *for* others, countering: "The implication of this celebration of place is that those of us who are not interested in place are less than alive, less than human, less than the sensitive geographers who are aware of such important things" (1993, 50). Rose equates the desire to interpret (things, places, etc.) with the desire to dominate and silence, urging readers to beware of this power play. They should instead adopt the tragic-ironic perspective that "interpretations" of the world and of places in it are signs. On this account, they would see attempts to communicate about the world (rather than about power) as assertions of the social order in the way that every sign rests entirely on consensus and, hence, on a kind of silence.

Rose's attention to Relph is indicative of the attention that feminist geography has directed toward questions of representation, the most sustained attention to communication issues within the discipline of geography. In *Feminism and Geography,* Rose states bluntly that "geography is masculinist" (1993, 4). She believes that a dissection of the languages of geography (both verbal and graphic) reveals that they constantly and chronically overlook the presence and the actions of women: although the discipline claims to be exhaustive, it in fact relates only to the position of men in the world, the geography of men. She not only critiques geography as masculinist but also takes the rhetorical approach of offering her book as "partial and strategic," two passwords meant to distance herself from what she sees as a male conceit: trying to present the whole, complete, and objective truth. Her circumscription of her own project and her condemnation of the entire body of geography demonstrate a profoundly tragic perspective on communication. This perspective is coupled with an ironic gesture: "I have attempted to take this text beyond the closures of masculinism's exhaustiveness through a movement between different analytical positions and a refusal to advocate one as better than the other" (1993, 15). This gesture is not comedy because comedy assumes that divergent interests can be reconciled, whereas Rose suggests only that different analytical positions should coexist in a single discussion. Her interest in maintaining several perspectives at once, including her own view, without resolution (closure), is the mark of irony and is quite prevalent in feminist geography. In short, Rose's tragic-ironic perspective denies the possibility of a common ground for discussion. People with different socially ascribed positions can say nothing to each other about the world that is not in fact a disguised attack. By seeming to provide a connection between self and other, language masks the violence of its actual purpose, which is to separate people. No one can speak for women but women. No one can speak for X but X. We must celebrate self-interest and condemn those (like Relph) who believe that one need not always preach to the choir.

Yet many new cultural geographers do speak on behalf of groups to which they do not belong. Developing some of Jackson's ideas, Tim Cresswell provides an intriguing account of how what is symbolically threatening is portrayed as "out of place," captured by such metaphorical associations as dirt, weeds, plagues, and bodily secretions. The judgment of what is "out of place" as bad, evil, or disruptive—transforming position into morality—is precisely the dynamic we would expect in the domain of symbols, and Cresswell demonstrates again and again that human bodies are treated as symbols in the landscape. His works demonstrate the symbolic contamination associated with out-of-place people: those individuals and groups who do not respect the rules of spatial behavior imposed by dominant groups. As an illustration, Cresswell describes a U.S. Department of Justice project called "Weed and Seed," meant to crack down on criminals in certain neighborhoods, then institute crime- and drug-prevention programs in those places. Symbolizing certain persons as weeds, dirt, infection, trash, or garbage is a way of justifying the argument that those people need to be removed in order to "clean up" the environment. "Once an inner-city resident is understood to be a weed, he or she can be treated like one" (1997, 343; see also N. Smith 1992, 1993).

Cresswell also shows how references to dirt may invert and disguise objectively defined health hazards. Women protesting at Greenham Common in Berkshire, England, were fighting a major hazard to health and life (a U.S. military installation with nuclear weapons), but those who opposed the women's presence cast *them* as a health hazard. The protesters' camp was symbolized in the language of dirt and excrement, making the out-of-place (both physically and politically) women, the camp, and their nonviolent, antimilitary agenda seem unnatural and morally repugnant, even monstrous.

Another permutation of the dirt metaphor can be found in conservative responses to the New Age Travellers, a contemporary hippie community in England. As the group gathered for festivals at ancient sacred places, its presence was described in terms of a "plague" threatening various forms of infection of local populations (Cresswell 1997, 337). The group's unusual spiritual and material attitudes, like the Greenham Common protesters' political and gender attitudes, were readily translated into a vocabulary of filth and infestation, which in turn justified "treatment" in the form of control, eviction, and expulsion.

In Cresswell's compelling account, what remains unexamined is how metaphors can (and do) serve to overcome rather than create social barriers. This oversight is committed by both Jackson and Rose as well. Particularly problematic is the assumption revealed by the quote with which Cresswell begins his article on "weeds": "A metaphor ... by virtue of what it hides, can lead to human degradation" (Lakoff and Johnson 1980, 236, quoted in Cresswell 1997, 330). He omits to mention that a metaphor, by what it reveals, can also lead to a better understanding between people and a clearer perception of the world. The comic interpretation I offer here complements the tragic interpretation of new cultural geography, symbolizing an equally real aspect of the extensible self. Cresswell's oversight is no doubt intentional at one level because he wants to make a point about a particular metaphor's political dimension. Yet he obscures (and therefore discourages) the possibility of engaging in communication's other dimensions, especially those that are constructive and creative rather than either oppressive or disruptive. His choice of examples is a selective vision revealing his cynical view of metaphor.

If we think of metaphors only as weapons entrained by the participants in social conflicts, we overlook other ways of thinking about metaphors: such as "metaphor as social glue" or "metaphor as a window on reality." Although Cresswell does not claim that people are destined in some lawlike fashion to use symbols against each other, he does assemble only those examples that demonstrate this strategic, combative use of symbolism while avoiding observations that would allow him to show how symbols overcome divisions and build new forms of social solidarity. This oversight of certain metaphorical symbols of metaphor itself is a perfect illustration of the bias of the tragic-ironic appeal.

The relationship between such views of communication and geography's *agenda* is illuminated by the writings of Gerard Toal (who has published under the Irish name Gearóid Ó Tuathail). Toal is a leading figure in the burgeoning subdiscipline called "critical geopolitics." He shares many rhetorical affinities with Jackson and Rose, and he would certainly support the general drift of Cresswell's characterization of metaphor. His approach to language is evident from the outset in *Critical Geopolitics* as he attacks the word *geopolitics* with a strategically aimed hyphen:

> Geo-politics does not mark a fixed presence but an unstable and indeterminate problematic; it is not an "is" but a question. The hyphen ruptures the givenness of geopolitics and opens up the seal of the bonding of the "geo" and "politics" to critical thought. In undoing the symbolic functioning of the sign, its semantic instability, ambiguity, and indeterminacy are released. The sign lies open before us, a dis-

rupted unity in question, a sign of a textual weave involving geography and politics. (Ó Tuathail 1996, 67)

Beginning with the dual assumption that "Geography is about power" and that communication is fundamentally political in nature (and thus adopting an erroneous sign-based model), Toal strives to destabilize and disrupt the naturalism of existing geographical portrayals of the world (Ó Tuathail 1996, 1). This approach is most clearly shown in his critique of Halford Mackinder's pedagogical theory. Mackinder, whose "heartland" theory inspired a half century of U.S. foreign policy, also had opinions about geographical education. He opposed rote memorization and argued that involvement, participation, and multisensory experience—in particular, field studies and map reading—would help geography students learn how to see the world more clearly and realistically. Toal condemns this hands-on approach to geographic pedagogy as a hidden power play: the claim to teach students techniques for perceiving the world promotes an insidious illusion that one person can help another see the real world more clearly. In light of Jacques Derrida's concept of "logocentrism," this claim implies "the dependence of theories of thought, discourse, or in this case, meaning on a metaphysical authority (Logos) that is considered external to them and whose truth and validity they express" (Ó Tuathail 1996, 65; a similar argument is advanced in Harley 1988, 1996). On this account, participatory forms of geographical education, such as field studies, are only superficially different from rote memorization; both amount to indoctrination, although one is more subtle. Instruction in map reading and field studies are in fact more insidious than rote memorization as vehicles for indoctrination because they produce the illusion of self-guided perception.

The apparently innocent claim that a geographer should be a person "of trained imagination" who can "analyze an environment" as the "local resultant of worldwide systems" is laid bare as a mechanism for reproducing the colonial imagination. Toal discerns "a romanticization of disciplining, a program of regulated training that is all the more pernicious for its thinking of itself as naturalistic" (1996, 104). The problem runs very deep and is based, he believes, on the assumption that maps and other representations are motivated by a correspondence to the objective world (a correspondence that I have, in contrast, defended on the basis that texts, although made up of signs and therefore arbitrary in their elements, are assembled in ways that do in fact have a testable, objectively grounded correspondence to certain aspects of the world). Furthermore, he asserts, the idea that one can be taught to see for oneself is a lie because seeing "for oneself" in fact means adopting a preprogrammed way of seeing based on oppressive dichotomies.

> To hold to the possibility of "thinking in images" or "thinking visually" presupposes a system of codes that allows one to designate certain objects as "images," "visual pictures," or "words." To hold to an epistemological stance structured around the innocent perception of external objects requires that one already operate with an epistemology that differentiates between the self and the world, consciousness and objects, an inside and an outside, passive and active, subject and object, and so on. (Ó Tuathail 1996, 105)

Insofar as colonialism depends on thinking in terms of these dichotomies, Toal believes, its worldviews must be unsettled. What is required is a worldview lacking all such distinctions. Toal's target is much broader than Mackinder because the vast majority of geography professors would support the goal of teaching students how to read and interpret maps, or at least to distinguish between subject/self and object/world. His critique clearly extends further than the critiques of the other authors mentioned here. Whereas Rose, Cresswell, and Jackson indicate a radical reformulation of geography, Toal is ultimately dismissing the moral defensibility of the entire project of learning and teaching about the nature of the world. The *pragmatic* implication of "logos" is that one simply cannot communicate about people or places without doing violence to those people and places—silence is a political and moral obligation.

Geographer Don Mitchell has applied many of the ideas laid out thus far to the task of defining cultural geography. He performs the ironic feat of writing a book about culture although, according to another article (1995), he does not believe in the existence of culture. In his own words, "the idea of culture is always an idea that works *for* some set of social actors (even if they are always opposed by other actors). The implementation of the idea of culture is a socially intentional process. The idea of culture is itself ideology" (2000, 78).

So his book is not about cultural geography but about "cultural" "geography," as captured in J. B. Thompson's description: "a system of signification which facilitates the pursuit of particular interests" (D. Mitchell 2000, 78). In other words, "culture" is something that people assert over and against other views of "culture" in order to advance the interests of their own group and to exploit or dominate other groups, and that act of domination or resistance is all there is to culture.

The following statement is key: "Culture and politics simply cannot be divorced, and efforts to do so merely play into the hands of those who would use 'culture' to exclude rather than include" (D. Mitchell 2000, 33 n. 39). What Mitchell means by this is that culture and politics are exactly the same: sexuality is politics, music is politics, literature is politics, daily life is politics, and, most important, landscapes are politics. In framing his argument, he distances himself from the traditional base-superstructure model of Marxism (and Jackson's early view), replacing ideology's unidirectionality with the bidirectionality of struggle and contestation: "cultural construction is not simply a top-down process; rather it is organized out of the seemingly infinite variety of 'resistances' in which people engage so as to maintain some semblance of control over the conditions of their own lives" (2000, 147–48).

Mitchell borrows the term *culture wars* from Patrick Buchanan (the far right and the far left often share metaphors), and the term becomes central to his argument that culture, politics, and economics are inseparable. This term and the tragic-ironic model behind it leads him to ridicule geographers who suggest that culture is essentially about differences between groups that may or may not engender conflicts (2000, 63–65). For Mitchell, difference between groups (as defined by culture) must engender conflict and has no meaning except as a basis for struggle. His world is a grim place indeed.

Wilbur Zelinsky's *Cultural Geography of the United States* (1973) comes under fire from Mitchell as "idealist" and "assimilationist" simply because Zelinsky includes (in his list of three aspects of culture) the idea that a culture is more than the sum of its parts. Zelinsky clears room for some part of culture to stand above the internal conflicts between subcultures. Oddly, Mitchell reorganizes Zelinsky's list to make this issue appear more central to Zelinsky's argument than it is. His harsh treatment of Zelinsky is motivated by the awareness that if a culture is more than the sum of its parts, then it is not *entirely* political because politics is restricted in scope to the interactions of the parts (as defined by class, ethnicity, gender, and so on). To understand why including something more than politics in culture is so offensive to Mitchell we must recognize that if winning and losing fails to describe culture, no matter how we cut up society into groups, then Mitchell's exclusively tragic explanation begins to look incomplete. The idea of "a culture" as containing some shared elements (as well as some differences) is an idea that incorporates both comic and tragic elements. Like all good orators, Mitchell wants to eradicate ambiguity, and the aesthetic perfection of his "culture war" image (a symbol) is disrupted by the (out of place) suggestion that certain things may not be fully explained in terms of oppression and resistance.

Clearly, Mitchell's view is incomplete because it (like the other views I have discussed) fails to explain how it is that people ever manage to transcend a particular conflict and work together, play together, or take part in creative activities together. We might imagine people cooperating with each other grimly because they calculate that "resistance is futile," but then where would the joy and delight come from when people participate in a collective activity such as a festival or parade?[6] From the sense of having left out some other group? By denying the existence of the romantic and comic feeling tone, Mitchell eliminates the worldviews that provide an antidote to the "culture war" as a social order, so his ver-

6. The grim phrase "resistance is futile" is of course the greeting of the evil Borg (cyborgs) in the television series *Star Trek: The Next Generation* and *Star Trek: Voyager.*

sion of radical politics is designed to prevent constructive, creative, mutually supportive ways of interrelating. His radical politics leads to paralysis, which is the opposite of its intended effect.

I end this exposition of geographical views of communication with David Harvey, a tremendously influential geographer. He is famous for, among other things, making a move from positivism to structuralism, but this was not as large a reorientation as is commonly thought.[7] In both cases, his focus has been on the constraints that hem in human action from all sides and on the individual's helplessness to cause significant change. Harvey has been a remarkably consistent adherent of the tragic narrative appeal and of the limited scope of human agency (with the only notable exception being a weakly developed postcapitalist fantasy in *Spaces of Hope* [2000]), even while swimming with the tide and following shifts in disciplinary interest from positivism to structuralism. In "Between Space and Time: Reflections on the Geographical Imagination," he argues that "the social definitions of space and time operate with the full force of objective facts to which all individuals and institutions necessarily respond" (1990, 418). This argument inverts the relationship between symbol and symbolizer, causing us to question which comes first. Perhaps symbols are stronger than the people who use them, and they serve to condition people to live in ways appropriate to their society's mode of production. To show that this relationship applies to manual laborers and intellectuals alike, Harvey borrows the term *cultural mass* from Daniel Bell to indicate artists, designers, and intellectuals as well. The "cultural mass" includes virtually all those persons with the level of education that would lead them to read his article. Harvey unflatteringly points out that they could not possibly achieve anything constructive because they are the most faithful servants of the system (1993, 25).

Whereas Harvey portrays educated people as a "mass," with overtones of lumpish immobility, he animates the invisible social formation with life: "in the face of a fierce bout of time-space compression, and of

7. For the positivist phase, see Harvey 1969; for the structuralist phase, see Harvey 1973, 1982, 1985, 1989, and 1990.

all the restructurings to which we have been exposed these last few years, . . . the security of place has been threatened and the map of the world rejigged as part of a desperate speculative gamble to keep the accumulation of capital on track" (1993, 27). People act as a mass, animated by the symbols of power that infect them, and thereby fight for capitalism even as its survival entails complete subjugation and mental submission. Yet Harvey's use of terms such as *rejigged* and *desperate* offers some hope; capitalism is teetering on the brink of collapse, he wants to suggest, and even if we are helpless to overthrow or improve it, we may soon enjoy its demise as a kind of grace descending on us. Rather than building connections based on trust, love, and creativity, people are expected more passively to resonate with a sense of millennial expectancy. How he, the author, avoids the plight of the rest of the "cultural mass" is not clear. He wants to evoke a sense of shock and awe in his audience so that they will wake up and change society, but his disingenuous argument instead inspires them to be either hypocritical or apathetic.

That the tragic-ironic mode is wedded to an organic view of society derived from materialism is less important in understanding Harvey than this rhetorical approach. What he argues for is withdrawal from communication and adoption of a distant, ironic gaze like that of social theorists, which has much in common with the ascetic worldview of medieval monastic orders.[8] Harvey's project seeks to uproot comic desires for community and to supplant them with tragic and ironic sensibilities of social struggle. Yet to join in the struggle implies condemning communication itself and revealing its impotence in the face of economic forces. Harvey is a modernist, whereas the new cultural geography positions discussed earlier are postmodern in outlook, but his tragic-ironic (in a word, cynical) view of communication is virtually identical to those positions.

▲ ▼ ▲

I plan to indicate what can be done from a geographical viewpoint with a more complete array of narrative appeals in hand, but first I men-

8. For some of the earliest works in this genre, see Peet 1977.

tion two moral concerns I have with the tragic-ironic perspective I have taken pains to elucidate. First, in denying the comic and romantic aspects of communication or in actively trying to undermine them—for example, by using *romantic* as a pejorative term—a large group of geographers has inspired in quite a few students and readers a distrust of communication itself. To write essays and research in this tradition requires the Orwellian technique of "doublethink"—espousing mutually contradictory ideas. This can at best serve progressive political goals in a delayed and equivocal fashion.

Second, in conflating symbols with signs, these authors have made it almost impossible for geographers to discuss any grounding for communication that is not social. A geographer who writes of something being "true" or "untrue," "real" or "artificial," "natural" or "unnatural" now invites ridicule from his or her colleagues who are wise enough to recognize that these terms have no referents and hence no objective validity, and furthermore that the terms are tools for social oppression so that any use of them is morally reprehensible. My point is that although these terms are constructions, that does not make them arbitrary, meaningless, or useless. Nor does it mean that their use is *primarily* about setting one group against or above another. To argue thus is not only to risk internal contradiction but to imply that silence has the moral high ground.

The problems human geographers identify with regard to comic and romantic appeals, on the one hand, and to symbols, on the other, are resolved if we identify the main social or communicational problem as residing not in communication content but in its contexts. When these contexts are exclusionary and rigidly, oppressively hierarchical, then the comedy becomes a farce, the romance an illusion, and symbols nothing but icons, like the cross or the flag. To symbolize the world, however, requires a sufficiently open-ended context of communication that symbols remain fluid and subject to revision, and that all four narrative appeals—tragic, comic, romantic, and ironic—are present and appreciated. As James Clifford observes, "There is no master narrative that can reconcile the tragic and comic plots of global cultural history" (1988,

15). This means that a tragic-ironic view of communication cannot be subjectively valid (sincere) if expressed because such expression presupposes the utility of communication and hence adopts a comic or romantic belief in reaching across social boundaries to exchange insights with members of one's audience. This submerged comic-romantic narrative should be brought to the surface through the recognition of *community,* as the expression of place by means of reconciliation and as the expression of reconciliation by means of place, and through various aspects of community: trust, love, and altruism in particular.

So who, among geographers, can serve as a better guide to the meaning of communication? Yi-Fu Tuan is the ideal guide here because his peculiar style is marked by the mingling of all four narrative appeals.

Comedy, Romance, and Emancipatory Worldviews

Tuan, in his not entirely uplifting autobiography, provides an account of a moment in his life, perhaps a momentary identity crisis. He characteristically roots a state of mind in a particular place and blends various narrative appeals to encapsulate the complexity of the experience of being in place.

> David must have decided that he knew the way and that there was no point in following me. Moreover, the children probably needed attention. I guessed these reasons. What happened was a beep from behind, then a rush of air as David's car overtook mine in a cloud of dust. I didn't want to be left too far behind, so I pressed down the pedal. But my truck was not made for speed. . . . The distance between us steadily widened. After a while, all I could see was the cloud of dust. Then, not even that. I was entirely alone in the deeply shadowed landscape. . . .
>
> As I strove to catch up with the Harrises, I felt alone for the first time—bitterly alone. Worse, I felt ridiculous. What on earth was I doing on that dirt road in New Mexico, chasing after a young family in its family car? I thought of turning the truck around and driving right back to Albuquerque. But of course, I didn't. Sanity returned. At the

motel in Cuba [New Mexico], I found that the Harrises had already changed the baby, who was cooing in her cot. We were all ready for something to eat. (1999, 50–52)

Tuan weaves together in this short passage the loneliness of a dirt road in the wilderness and the loneliness of a single man trailing after a friend who has a family. The fact that the author is gay and we are not sure of his feelings toward David Harris adds a layer of ambiguity that is not entirely resolved, but blends with the mysterious quality of the darkened desert landscape. Two vehicles, two speeds, two lifestyles, and two senses of place contribute to the image. Although this artful passage says little about geography and much more about a turning point in a young man's life, I have chosen it to start my discussion of blending appeals because the comic appeal is not applied indiscriminately in it; rather, the comic is pulled out of the hat at the end. What Tuan really missed, it turns out, was not just companionship, on various levels, but the changing of a dirty diaper in the motel. The loneliness of the empty road is swallowed up by the final act: coming together over a meal. Tuan's skill lies largely in his ability to move gracefully from dark to light and vice versa, revealing the complexity of human experience and ultimately the depth and subtlety of place.[9]

Tuan argued early in his career (Tuan 1978) that geographers benefit from literature in three ways: (1) as a stylistic example worthy of emulation, (2) as a source of information about other times and places, and (3) as a means of insight into personal experiences of the world. The "stylistic" inspiration of literature suggests that he is sensitive to all four of the narrative appeals. Indeed, he uses them in a balanced fashion, as if committed to the idea that an accurate portrayal of being-in-the-world requires the poignancy of mixtures such as the tragic-comic and the romantic-ironic. We must recall Hayden White's argument that such

9. This rhetorical flexibility may stem in part from the author's lifepath. At age eleven, Tuan moved with his family from China to Australia, a linguistic and cultural boundary crossing. If language forms the substrate of experience, then surely bilingualism and biculturalism can inspire unusual breadth of thought, which can in turn be reflected in rhetorical flexibility.

appeals are not just packaging for facts, but complementary aspects of objective reality, though not all geographers are as talented with the English language as Tuan.

In Tuan's descriptions of place, we find evidence that small things—newly planted trees, streetlamps, patches of earth worn bare by the passage of feet, even motel rooms—can be subjectively "touching," reassuring, and positive experiences of place. The literary passages he uses to support this insight are comic in character: people waving to friends, the author admiring a sunset or nostalgically recalling childhood. (The literary works he draws on were written in the 1960s and 1970s, before the popular style shifted to favor irony.) Tuan seldom, at least in his mature works, arrives at comic reconciliation without pointing out the tragic divides that separate people.

Throughout his writings, Tuan has worked to discover universal qualities of the human experience of place and space. His approach often dwells on particular communications—quotes, observations, historical accounts, and anthropological findings. He takes delight in communication—his own communication and others'—in a way that can easily be denounced as romantic. Indeed, in his search for the universal in the particular, he aligns himself explicitly with the romantic project of transcending culturally specific worldviews through cross-cultural understanding and the comic aim of achieving a moral and a *good life* (Tuan 1986a). His effort to raise the question of the good life has met with silence from geographers, as it is perhaps considered too philosophical (or simply too romantic), yet it completes the cycle begun by critique, gesturing toward subjective and intersubjective responses to problems such as racial bias, class inequality, sexism, and homophobia. Tuan reveals communication as a mechanism of dominance at times, but he refuses to leave love, hope, imagination, creativity, aesthetics, and spirituality out of his accounts of the world. He recognizes that the comic and romantic appeals are essential to those who want not only to critique, but to inspire, not only to tear apart, but to create (Tuan 1984).

In Tuan's attention to creativity and human attachments, we can therefore sense the comic sensibility as well as the romantic (Tuan 1974, 1988, 1996). Yet in his interest in escapism, dominance, isolation, punishment, and deception, he also demonstrates a mastery of the tragic and ironic sensibilities (Tuan 1979, 1998). Most of his later works blend these modes, much to the consternation of geographers who prefer a simpler diet (Tuan 1977, 1982, 1986a).

It is telling that his autobiography *Who Am I?* (1999) sounds the note of tragedy most often; he does not hesitate to admit his shortcomings, weaknesses, and failures and appears almost to offer the book as a challenge to those who see the world as a tragedy but who clandestinely think of their own academic and personal careers in romantic terms. To present one's life as a tragedy demonstrates a kind of generosity because the author's failures suggest to the reader that he or she can be at least as successful, if success is properly defined. Lurking behind the tragic autobiography is a discourse in which the great life dissolves to leave the good life transcendent. Again, Tuan has pulled the comic appeal out of the hat by rejecting the opportunity to glorify his academic career.

An extensive study of communication can be found in the chapter on theater in Tuan's *Segmented Worlds and Self* (1982). He finds an intriguing parallel between, on the one hand, the act of bounding and isolating a place for performance by moving theater from the street to a well-defined stage, which took place in Western culture over the past four hundred years, and, on the other, the shift in dramatic subjects from grand cosmological themes to the secret spaces of private relationships, the inner space of the household, and, beyond that, the most mysterious world of the psyche. He suggests that behind this reorganization of a particular kind of place lies a more general social and spatial shift from the premodern world with its unified worlds and selves to the modern world of segmented worlds and selves. Ever sensitive to the interweaving of tragic and comic elements, he argues: "Deep personal relationships presuppose the existence of persons, that is, complex and self-aware individuals; but such individuals can emerge only as the cohesive and unreflective nature of community begins to break down" (1982, 196). So an interior space, even a neurotic interior space in a fragmented world (such as the living rooms in plays by Edward Albee [1988]) is one of the preconditions of individuation and hence the necessary adjunct of a more inquisitive and intellectually unbounded relationship to the world. Modernity's prevalent forces of personal isolation have brought discontent to individuals and supported various oppressive and exploitative relations between people, but they at the same time provide a source of individual empowerment, helping people transcend narrowly defined conceptions of self and reconcile conflicting objectives and desires. Tuan always wraps up the tragic and the comic, the ironic and the romantic, in a single parcel and presents this parcel as a bittersweet glimpse of the wonderful world we inhabit.

When writing about communication, Tuan has more complex and nuanced things to say than any other geographer: "The pocket-size book was invented in the sixteenth century, and it has proven to be the most convenient and safest means of escapism known to humankind. With a book, we can be in the Arctic and escape to the tropical forest, in Grand Central Station and escape to Hawaii, in an office rest room and escape to a Bangkok brothel, in the year 2000 and escape to the year 1000" (2002, 165–66). Here we find recognition of virtual place couched in romantic terms, but a hint of corruption (pornography) suggests that within the romance of escape lies the tragedy of desires that can never be satisfied as well as the tragedy of exploitation.

The comic theme of mutual understanding is demonstrated in Tuan's appreciation of a confession. The confession comes from Charles Darwin with respect to being "a poor critic." Tuan writes:

> Darwin doesn't seem aware that he is describing an advantage rather than a handicap. Unlike his quick-witted friends who go into a book with darts all ready to fly and therefore can hardly learn anything from it, Darwin with his receptivity—his readiness to admire—can always pick up some useful facts or even an insight before applying the dismissive intellect. Charity, it turns out, is highly desirable even in operations of the mind. (2002, 107)

Here the comic tone—and more than this, a comic vision—is applied to Darwin. The account also exemplifies an attitude toward the act of reading and thence toward communication itself. Tuan implies that it is good, in a general sense, to temper the critical sensibility when interpreting the meaning of writing (and presumably other communications).

"Tuanian" writing perhaps means geographical writing that illuminates not just the subjective qualities of place, but also the comic and romantic elements of experience. Still, Tuan is no stranger to irony, which he employs in a more irreverent fashion than one expects from a geographer. *"Mist,"* he observes, "What a beautiful word in English, calling up, for me, the image of an English meadow early in the morning. But in German it means *crap!* I conclude that anyone with serious poetic ambition can't risk being bilingual" (2002, 170). Tuan points out the arbitrariness of the signifier-signified link in this example and does so in characteristically "Tuanian" fashion.

Tuan is not oblivious to the tragic side of place, of being in place, or of communication itself. Drawing from the Bible, he reflects: " 'What is truth?' Pilate asked. One who ought to know chose silence as his answer" (2002, 170). In an instant, he reminds us that one never ceases to communicate, yet communication in words is always short of the truth. He suggests as well that to make sense implies automatically a kind of oversight, which has discomforting moral implications. In short, he raises the stakes of critical communication theory by generating a view that is not only critical, but fluid with regard to the "space" defined by rhetorical appeals. This fluidity is not mere artistry, but a necessary foundation for a rich, complex, and nuanced portrayal of the human experience of place.

These last selections are from his collection of *Dear Colleague* letters, published in 2002 as he looked back on a long and illustrious career. The successful career is not unrelated to Tuan's understanding of communication. He is one of the handful of geographers who is well known outside the community of professional geographers, and his communications have been unusually well received. We might infer that he broadened his extensibility (via the printed word) through his grasp of various rhetorical appeals, in particular by weaving the overarching symbols of the comic, tragic, romantic, and ironic all together. This breadth leads to other kinds of breadth: social, moral, even geographical. For instance, his works have appeared in sixteen translated editions, permitting the conversion of rhetorical breadth into geographical breadth.

His most important contribution has been to situate communication as a fundamental element of the human-environment relationship. Tuan argues that "Language is important to students of place not only because a Thomas Hardy or Willa Cather has written evocatively on landscape, and has thus provided a literary standard that geographers should seek to emulate in their own writing; rather language is important—indeed central—because humans are language animals, and language is a force that all of us use everyday to build, sustain, and destroy" (1991, 694). He means this argument quite literally, as when a construction worker concretizes the spaces encoded in an architect's blueprint, but also symbolically, as when names on a map shape the way a place is explored, claimed, and modified. His observation knits together the place-creating powers of both content and context: when filled with conversation, a room takes on a particular quality and meaning; when brought to life in literary images, a place is experienced differently by those who encounter or inhabit it. He shows in countless ways that communication is contained in places, and places are contained in communication. In this paradoxical mutual enfolding of place and communication, the human individual grows and defines itself, reaching outward to engage with a wide world and often to express impulses of love for people and places, the willingness to help without an expectation of rewards, and the ability to trust in others, yet he or she is forever limited in this endeavor by fear, dominance, and ethnocentrism. Tuan shows us these latter elements, but offers a more optimistic vision as well.

The Lesson of Optimism

In contrast to many current accounts in the social sciences and geography in particular, I have tried to stress the way communication connects rather than the way it divides, the way it reflects the real world rather than the arbitrariness of code, the way social contexts include cooperation as well as conflict. Behind my endeavor is the model of the extensible individual, which I believe needs, for moral and practical reasons, to recognize its boundlessness. We must constantly bear in mind our connections to other persons and to the earth—connections that are not "things" we "have," but parts of self that merge with other selves. It serves no purpose to critique communication as a means of opposing and oppressing others or as a means of distinguishing oneself from them and assigning them inferior status if in the process we lose sight of the fact that this aspect of communication coexists with the connective aspect, or else communication would be impossible. The hope for communication lies in the comic image of connection, not in the tragic image of separation, because the tragic claims undermine the very form of agency that they wish to employ.

Communication must be seen as a moral imperative for individuals who exist in and through connections that defy the apparent insularity of self. My suggestion, then, is not to expose, destabilize, and deconstruct communication, but to suggest instead that communication remains above all a way of responding to shared concerns and even a way of responding to the concerns of persons who cannot join in our "conversation" because of their separation in time or distance. To make this claim I assume the existence of trust and altruism. An author can do little to inculcate these attitudes, but the attitudes in question are already there, as part of the "equipment" of extensible readers, and should not be disparaged in the name of a critical perspective that denies its own hopefulness and hence is strangely insincere.

When communication is seen only as "a struggle of adversaries who strive to usurp each other's meaning, who dodge and squirm to escape each other's dictations, who are simultaneously recalcitrant and imperious, who would define but will not be defined," trust and altruism cannot flourish.[10] To promote cooperation and nonviolence demands that we acknowledge the cooperative and collaborative elements of communication even as we work to expose the many sources of decep-

10. These words are Jonathan Smith's (1993a, 90) paraphrase of Harold Bloom's notion of reading.

tion and symbolic coercion in society. My concern with this point is at once pragmatic and moral, which seem rather disparate concerns, but which draw together if we consider communication over long periods of time.

All communications—subjective, objective, and intersubjective in nature—are locked in a feedback loop. We can not opt out of communicative action without ceasing to be human. For this reason, we must trust not only in each other but also in the languages (verbal, graphic, cartographic, mathematical, musical, etc.) that are at our disposal. Our trust is based on the cyclical nature of action in the world. Just as a hawk that fails to catch its rodent supper will try again and again until it succeeds, modifying its dive or swoop until the objective is met, so languages are reflexively altered as a vital part of human-environment interrelations. A human population, be it a village or the world community, will similarly learn from its encounters with the environment; it will modify its activities in response to pragmatic goals. The learning process occurs in signs, symbols, and signals, but it also involves the reworking of the systems of signs, symbols, and signals. A hawk that misses its prey is learning how to hunt, and it learns without words, at the levels of signals and symbols. Human groups learn also at the symbol and signal levels, but insofar as words situate people imaginatively in a broader space and time than the proximate environment of the body, words evolve as instruments for coordinating extensible individuals' actions. Signs themselves and the rules that govern their use are arbitrary. But symbols made of signs, such as scientific books, essays, and articles, are not arbitrary. Their construction reflects a process of trial and error, like the hawk's hunting skills, drawn out over generations and grounded in environments (both physical and social) through reflexive discussion.

I do not mean to support the untenable argument that culture itself has needs, wants, desires, or will. The sole motivators of this adaptive process are people, yet people exist in networks where members are defined by communications. Every communication act is therefore a kind of test within the virtual environment of one's social connections. What constitutes a "bad" message can be characterized by a range of validity problems, from falsity to insincerity to injustice. In various ways, each of these kinds of error is a failure to represent the world—social or natural or both—in mediated form. New technologies also mediate, and what makes them *new* is that they mediate between people and the world in new ways, with new topologies of connection and new combinations of signs, symbols, and signals. We can easily overstate the difference between old and new media, however, and must take care to focus our analysis on the more persistent social and personal roles of communication. New media are caught up in the same cycle of communicative action and perceived consequences that has existed since the verbalization of the first signs by our hominid ancestors. New media are not simply extensions of the capitalist commodity system or of instrumental rationality. Instead, they are environments in which the symbolic representation of the world is continued and refined (that refinement now including the presence of new communication infrastructure as both a subject of discussion and a means of forming communication links).

If our communications include within them the primary message that communication itself is not to be trusted (as is presently the case with critical theory in geography), then certain individuals will be inclined to detach themselves as much as possible, at least on the emotional level, from the process of representing the world. Such disassociation responds to foundational assumptions: *(a)* that communication consists of sign systems (at the level of content), and *(b)* that representations necessarily recapitulate social hierarchies (at the level of context) so that one can build an argument that every communication act is a kind of power, even an act of violence, imposed on the other. To break with this critical tradition is to risk being branded as *romantic*. Properly understood, the *r* word is appropriate; it is even a compliment. Romantic narratives are exactly what is needed to reinfuse academic communication with a sense of possibility and to encourage greater personal autonomy.

The questions remain, however: *What is* the pragmatic purpose of romance, what is the function of comedy, and, in the most general terms, what is the lesson of optimism? When human society changes, or when human-environment relations change, certain people are involved in that change. They do not single-handedly cause the change, but they play a vital role in it. Let us presume, for example, that in the future some sort of measures will be taken to reduce the emission of greenhouse gasses into the earth's atmosphere. Participation in such a change will almost certainly depend on participants having a sense of self that includes an attachment to the motifs of transcendence and reconciliation. Current ways of producing energy, of arranging the built environment, and of getting from place to place will have to be transcended because they are based on forms of dominance and separation. How can those who do not believe in the transcendence of their present conditions become active in changing people's habits and beliefs? How can those who do not believe in reconciliation take part in a debate that seeks to build consensus? Romantic and comic narratives are crucial media for promoting politically progressive constructions of the individual because they support a certain willingness to engage with others on behalf of a change.

The necessity of romantic and comic narratives increases as the individual senses himself or herself to be operating "in the dark" as a result of living in a distanciated society. Stated simply, extensible individuals can live moral lives only if they are able to internalize symbols of transcendence and reconciliation. Optimistic rhetoric promotes awareness of extensibility, which in turn permits the rebalancing of spaces of communicative action with spaces of impersonal, system-assisted action (such as consumption). Ultimately, then, I believe rhetoric holds the key (or at least a key) to reversing the colonization of the lifeworld. Other solutions may reside at the organizational level, but again I might argue that a comic or romantic personal stance balanced with a measure of cynicism is required for people to participate constructively in such organizations.

It would be folly to adopt a rhetorical approach of pure romance or pure comedy. Just as change is real, so is stasis. Just as reconciliation is real, so is conflict. In the words of Gunnar Olsson, "the only false perspective is the one which pretends to be unique" (1975, 29).

The world becomes known to people (and to animals) through the constraints it imposes on actions. Whether natural or social in origins, these limitations are symbolized through tragic narratives. Tragedy in-

corporated in worldviews and actions helps people to predict and therefore to avoid hazards, tensions, struggles, setbacks, and other difficulties. Animals would probably tell us tragic tales if they could talk
because tragedy certainly is the motif most suited to their place-bound
and time-bound existence. And the tragic appeal is pragmatically useful
to people insofar as they need to adapt to surroundings more often than
they are able to change them.

However, the distanciated world both constrains and enables our
actions. Enablement cannot be expressed in tragic terms because its
essence is agency, which tragic narratives deny. If we weave our portrayals of the world broadly, in the symbolic fabric of tragedy and comedy,
romance and farce, then we are providing a wealth of resources for active, realistic, and even satisfying engagement with the world as a progressive force.

On a final note, how can we possibly control the way our optimistic
appeals are *used?* Our romantic tale might inspire an oppressor cheer-
fully to expand his or her scope of control, and our tragedy might push
someone who is oppressed to despair. The simple fact is that we cannot
control this aspect of communication. As I argued at the outset, when I
speak and you listen, something is exchanged, but, more important,
something is shared that allows my words to be recognized, not only as
a collection of signs, hollow and arbitrary as ciphers, but as an experience of being-in-the-world. That "something" is a mystery I must treat
with caution but also with trust.

References

Abler, Ronald. 1975a. "Effects of Space-Adjusting Technologies on the Human Geography of the Future." In *Human Geography in a Shrinking World,* edited by Ronald Abler, Donald Janelle, Allen Philbrick, and John Sommer, 36–56. North Scituate, Mass.: Duxbury Press.

———. 1975b. "Monoculture or Miniculture: The Impact of Communications Media on Culture in Space." In *Human Geography in a Shrinking World,* edited by Ronald Abler, Donald Janelle, Allen Philbrick, and John Sommer, 122–48. North Scituate, Mass.: Duxbury Press.

———. 1977. "The Telephone and the Evolution of the American Metropolitan System." In *The Social Impact of the Telephone,* edited by Ithiel de Sola Pool, 318–41. Cambridge, Mass.: MIT Press.

Adamic, L. A. 1999. "The Small-World Web." In *Proceedings of the European Conference on Digital Libraries 1999 Conference,* 443–52. Berlin: Springer.

Adams, Paul. 1992. "Television as Gathering Place." *Annals of the Association of American Geographers* 82, no. 1: 117–35.

———. 1995. "A Reconsideration of Personal Boundaries in Space-Time." *Annals of the Association of American Geographers* 85, no. 2: 267–85.

———. 1996. "Protest and the Scale Politics of Telecommunications." *Political Geography* 15, no. 5: 419–41.

———. 1997. "Computer Networks and Virtual Place Metaphors." *Geographical Review* 87, no. 2: 155–71. [Special Issue on Cyberspace and Geographical Space. Guest editors, P. Adams and B. Warf.]

———. 1998. "Network Topologies and Virtual Place." *Annals of the Association of American Geographers* 88, no. 1: 88–106.

————. 1999. "Bringing Globalization Home: A Homeworker in the Information Age." *Urban Geography* 20, no. 4 (May 16–June 30): 356–76.

————. 2000. "Application of a CAD-Based Accessibility Model." In *Information, Place, and Cyberspace: Issues in Accessibility,* edited by Donald Janelle and David Hodge, 217–39. New York: Springer.

————. 2001. "Peripatetic Imagery and Peripatetic Sense of Place." In *Textures of Place: Exploring Humanist Geographies,* edited by Paul C. Adams, Steven Hoelscher, and Karen Till, 186–206. Minneapolis: Univ. of Minnesota Press.

Adams, Paul, and Barney Warf. 1997. "Introduction: Cyberspace and Geographical Space." *Journal of Geography* 87: 139–45.

Adorno, Theodore. 1981. *Prisms.* Translated by Samuel Weber and Shierry Weber. Cambridge, Mass.: MIT Press.

Ainley, Rosa. 1998. *New Frontiers of Space, Bodies, and Gender.* New York: Routledge.

Albee, Edward. 1988. *Who's Afraid of Virginia Woolf?* New York: New American Library.

Albert, Réka, Hawoong Jeong, and Albert-Laszlo Barabási. 1999. "Diameter of the World Wide Web." *Nature* 401: 509–12.

Alway, Joan. 1995. *Critical Theory and Political Possibilities: Conceptions of Emancipatory Politics in the Works of Horkheimer, Adorno, Marcuse, and Habermas.* Westport, Conn.: Greenwood Press.

Anderson, Benedict. 1983. *Imagined Communities.* New York: Verso.

Appadurai, Arjun. 1996. *Modernity at Large: Cultural Dimensions of Globalization.* Minneapolis: Univ. of Minnesota Press.

Bachelard, Gaston. 1969. *The Poetics of Space.* Translated by Maria Jolas. Boston: Beacon Press.

Balnaves, Mark, James Donald, and Stephanie Hemelryk Donald. 2001. *The Penguin Atlas of Media and Information: Key Issues and Global Trends.* New York: Penguin Putnam.

Barabási, Albert-Laszlo. 2002. *Linked.* New York: Penguin.

Barnes, Trevor, and James Duncan. 1992. "Introduction." In *Writing Worlds: Discourse, Text, and Metaphor in the Representation of Landscape,* edited by Trevor Barnes and James Duncan, 1–17. New York: Routledge.

Barthes, Roland. 1967. *Elements of Semiology.* Translated by Annette Lavers and Colin Smith. New York: Hill and Wang.

————. 1972. *Mythologies.* Selected and translated by Annette Lavers. New York: Hill and Wang.

————. 1979. *The Eiffel Tower and Other Mythologies.* Translated by Richard Howard. Berkeley and Los Angeles: Univ. of California Press.

Baudrillard, Jean. 1983. *Simulations.* Translated by Paul Foss, Paul Patton, and Philip Beitchman. Foreign Agents series. New York: Semiotext(e).

Bell, David, and Gill Valentine. 1995. *Mapping Desire: Geographies of Sexualities.* New York: Routledge.

Benjamin, Walter. [1969] 1986. "The Work of Art in the Age of Mechanical Reproduction." Reprinted in *Video Culture: A Critical Investigation,* edited by J. Hanhardt, 27–52. Layton, Utah: Peregrine Smith.

Bianculli, David. 1992. *Teleliteracy: Taking Television Seriously.* New York: Continuum.

Bird, Jon, Barry Curtis, Tim Putnam, George Robertson, and Lisa Tickner, eds. 1993. *Mapping the Futures: Local Cultures, Global Change.* New York: Routledge.

Boal, Iain A. 1995. "A Flow of Monsters: Luddism and Virtual Technologies." In *Resisting the Virtual Life: The Culture and Politics of Information,* edited by James Brook and Iain A. Boal, 3–15. San Francisco: City Lights.

Bolter, David J. 1984. *Turing's Man: Western Culture in the Computer Age.* Chapel Hill: Univ. of North Carolina Press.

Bonnifield, Mathew Paul. 1979. *The Dust Bowl: Men, Dirt, and Depression.* Albuquerque: Univ. of New Mexico Press.

Borges, Jorge Luis. 1988. "The Garden of Forking Paths." In *Labyrinths: Selected Stories and Other Writings,* 19–29. New York: New Directions.

Bormann, F. Herbert, Diana Balmori, and Gordon T. Geballe. 1993. *Redesigning the American Lawn: A Search for Environmental Harmony.* Edited and researched by Lisa Bernegaard. New Haven, Conn.: Yale Univ. Press.

Bourdieu, Pierre. 1984. *Distinction: A Social Critique of the Judgement of Taste.* Translated by Richard Nice. Cambridge, Mass.: Harvard Univ. Press.

Bronner, Stephen. 1998. "Dialectics at a Standstill: A Methodological Inquiry into the Philosophy of Theodor W. Adorno." Section One. Available at: www.uta.edu/huma/illuminations/bron2.htm).

Brooks, David. 2000. *Bobos in Paradise: The New Upper Class and How They Got There.* New York: Simon and Schuster.

Buttimer, Anne. 1982. "Musing on Helicon: Root Metaphors and Geography." *Geografiska Annaler* Series B, 64: 89–96.

Calvino, Italo. 1974. *Invisible Cities.* Translated by William Weaver. San Diego: Harvest and Harcourt Brace Jovanovich.

Carey, James. 1989. *Communication as Culture: Essays on Media and Society.* Boston: Unwin Hyman.

Castells, Manuel. 1996. *The Rise of the Network Society.* Vol. 1 of *The Information Age: Economy, Society, and Culture.* Oxford: Blackwell.

———. 1997. *The Power of Identity.* Vol. 2 of *The Information Age: Economy, Society, and Culture.* Oxford: Blackwell.

———. 1998. *End of Millennium.* Vol. 3 of *The Information Age: Economy, Society, and Culture.* Oxford: Blackwell.

———. 1999. "Grassrooting the Space of Flows." *Urban Geography* 20, no. 4: 294–302.

Certeau, Michel de. 1984. *The Practice of Everyday Life.* Translated by Steven Rendall. Berkeley and Los Angeles: Univ. of California Press.

Chambers, J. K. 1995. *Sociolinguistic Theory: Linguistic Variation and Its Social Significance.* Oxford: Blackwell.

Clifford, James. 1988. *The Predicament of Culture: Twentieth-Century Ethnography, Literature, and Art.* Cambridge, Mass.: Harvard Univ. Press.

Coleman, James S. 2000. "Social Capital in the Creation of Human Capital." In *Social Capital: A Multifaceted Perspective,* edited by Partha Dasgupta and Ismail Serageldin, 16–17. Washington, D.C.: World Bank.

Cosgrove, Denis E. 1984. *Social Formation and Symbolic Landscape.* With a new introduction. Madison: Univ. of Wisconsin Press.

———. 2001. "Geography's Cosmos: The Dream and the Whole Round Earth." In *Textures of Place: Exploring Humanist Geographies,* edited by Paul C. Adams, Steven Hoelscher, and Karen E. Till, 326–39. Minneapolis: Univ. of Minnesota Press.

Cosgrove, Denis, and Stephen Daniels, eds. 1988. *The Iconography of Landscape.* Cambridge: Cambridge Univ. Press.

Couclelis, Helen. 1988. "The Truth Seekers: Geographers in Search of the Human World." In *A Ground for Common Search,* edited by Reginald Golledge, Helen Couclelis, and Peter Gould, 148–55. Goleta, Calif.: Santa Barbara Geographical Press.

Cresswell, Tim. 1996. *In Place/Out of Place: Geography, Ideology, and Transgression.* Minneapolis: Univ. of Minnesota Press.

———. 1997. "Weeds, Plagues, and Bodily Secretions: A Geographical Interpretation of Metaphors of Displacement." *Annals of the Association of American Geographers* 87, no. 2: 330–45.

———. 2001. "The Making of the Tramp." In *Textures of Place: Exploring Humanist Geographies,* edited by Paul C. Adams, Steven Hoelscher, and Karen Till, 167–85. Minneapolis: Univ. of Minnesota Press.

Crossan, John Dominic. 1994. *Jesus: A Revolutionary Biography.* San Francisco: HarperSanFrancisco.

Cruz, Donna de la. 2003. "During Blackout, Fewer Crimes Than on a Normal NYPD Day." Associated Press, available at: www.newsday.com/news/local/wire/ny-bc-ny—blackout-crime0816aug16,0,3480478.story?coll=ny-ap-regional-wire. Accessed Aug. 16, 2003.

Curry, Michael. 1996. *The Work in the World: Geographical Practice and the Written Word.* Minneapolis: Univ. of Minnesota Press.

———. 1997. "The Digital Individual and the Private Realm." *Annals of the Association of American Geographers* 87, no. 4: 681–99.

Dasgupta, Partha, and Ismail Serageldin, eds. 2000. *Social Capital: A Multifaceted Perspective.* Washington, D.C.: World Bank.

Dasgupta, Shamita Das. 1989. "Gender Roles and Cultural Continuity in the Asian Indian Immigrant Community in the U.S." *Sex Roles* 38, nos. 11–12: 953–74.

Deleuze, Gilles, and Félix Guattari. 1983. *Anti-Oedipus: Capitalism and Schizophrenia.* Translated by Robert Hurley, Mark Seem, and Helen R. Lane. Minneapolis: Univ. of Minnesota Press.

Deregowski, Jan B. 1973. "Illusion and Culture." In *Illusion in Nature and Art,* edited by R. L. Gregory and E. H. Gombrich, 161–91. New York: Charles Scribner's Sons.

Doel, Marcus, and David B. Clarke. 1999. "Virtual Worlds: Simulation, Suppletion, S(ed)uction, and Simulacra." In *Virtual Geographies: Bodies, Space, and Relations,* edited by Mike Crang, Phil Crang, and Jon May, 261–83. New York: Routledge.

Dryzek, John. 1995. "Critical Theory as a Research Program." In *The Cambridge Companion to Habermas,* edited by Stephen K. White, 97–119. Cambridge: Cambridge Univ. Press.

Duncan, James. 1990. *The City as Text: The Politics of Landscape Interpretation in the Kandyan Kingdom.* Cambridge: Cambridge Univ. Press.

Duncan, James, and David Ley, eds. 1993. *Place/Culture/Representation.* New York: Routledge.

Durkheim, Émile. 1897. *Le suicide: Étude de sociologie.* Paris: Alcan.

Eco, Umberto. 1976. *Theory of Semiotics.* Bloomington: Indiana Univ. Press.

Eisenstein, Elizabeth. 1979. *The Printing Press as an Agent of Change: Communications and Cultural Transformations in Early-Modern Europe.* Cambridge: Cambridge Univ. Press.

Entrikin, J. Nicholas. 1991. *The Betweenness of Place: Towards a Geography of Modernity.* Baltimore: Johns Hopkins Univ. Press.

Fiske, John. 1987. *Television Culture.* London: Methuen.

Fiske, John, Bob Hodge, and Graeme Turner. 1987. *Myths of Oz: Reading Australian Popular Culture.* Boston: Allen and Unwin.

Foucault, Michel. 1979. *Discipline and Punish: The Birth of the Prison.* Translated by Alan Sheridan. New York: Vintage.

———. 1980. *Power/Knowledge: Selected Interviews and Other Writings 1972–1977.* Edited by Colin Gordon. Translated by Colin Gordon, Leo Marshall, John Mepham, and Kate Sopher. New York: Pantheon.

Fouts, Roger. 1997. *Next of Kin: What Chimpanzees Have Taught Me about Who We Are.* New York: William Morrow.

Fromm, Erich. 1956. *The Art of Loving.* New York: Harper.

Frye, Northrop. 1957. *Anatomy of Criticism.* Princeton, N.J.: Princeton Univ. Press.

Gerbner, George. 2002. *Against the Mainstream: The Selected Works of George Gerbner.* Edited by Michael Morgan. New York: P. Lang.

Gergen, Kenneth J. 1991. *The Saturated Self: Dilemmas of Identity in Contemporary Life.* New York: Basic.

Giddens, Anthony. 1984. *The Constitution of Society: Outline of the Theory of Structuration.* Berkeley and Los Angeles: Univ. of California Press.

———. 1991. *Modernity and Self-Identity: Self and Society in the Late Modern Age.* Stanford, Calif.: Stanford Univ. Press.

Goffman, Erving. 1959. *The Presentation of Self in Everyday Life.* Woodstock, N.Y.: Overlook Press.

Gottdiener, Marc, and Alexandros Ph. Lagopoulos, eds. 1986. *The City and the Sign: An Introduction to Urban Semiotics.* New York: Columbia Univ. Press.

Gregory, R. L., and E. H. Gombrich, eds. 1973. *Illusion in Nature and Art.* New York: Charles Scribner's Sons.

Habermas, Jürgen. 1971. *Knowledge and Human Interests.* Translated by Jeremy Shapiro. Boston: Beacon Press.

———. 1984. *Reason and the Rationalization of Society.* Vol. 1 of *The Theory of*

Communicative Action. Translated by Thomas McCarthy. Boston: Beacon Press.

———. 1987. *Lifeworld and System: A Critique of Functionalist Reason*. Vol. 2 of *The Theory of Communicative Action*. Translated by Thomas McCarthy. Boston: Beacon Press.

———. 1990. "Discourse Ethics: Notes on a Program of Philosophical Justification." In *Moral Consciousness and Communicative Action,* translated by Christian Lenhardt and Shierry Weber Nicholsen, 43–115. Cambridge, Mass.: MIT Press.

Hägerstrand, Torsten. 1970. "What about People in Regional Science?" Presidential address to the Ninth European Congress of the Regional Science Association. *Papers of the Regional Science Association* 24: 7–21.

———. 1982. "Diorama, Path, and Project." *Tijdschrift voor Economische en Sociale Geografie* 73: 323–39.

———. 1983. "In Search for the Sources of Concepts." In *The Practice of Geography,* edited by Anne Buttimer, 238–56. London: Longman.

Haraway, Donna. 1985. "A Manifesto for Cyborgs: Science, Technology, and Socialist Feminism in the 1980s." *Socialist Review* 15, no. 2 (Mar.–Apr.): 65–107.

Harley, J. B. 1988. "Maps, Knowledge, and Power." In *The Iconography of Landscape: Essays on the Symbolic Representation, Design, and Use of Past Environments,* edited by Denis Cosgrove and Stephen Daniels, 277–312. Cambridge: Cambridge Univ. Press.

———. 1996. "Deconstructing the Map." In *Human Geography: An Essential Anthology,* edited by John Agnew, David N. Livingstone, and Alisdair Rogers, 422–43. Oxford: Blackwell.

Harvey, David. 1969. *Explanation in Geography*. London: Edward Arnold.

———. 1973. *Social Justice and the City*. London: Edward Arnold.

———. 1982. *Limits to Capital*. Oxford: Blackwell.

———. 1985. "The Geopolitics of Capitalism." In *Social Relations and Spatial Structures,* edited by Derek Gregory and John Urry, 128–63. London: Macmillan.

———. 1989. *The Condition of Postmodernity: An Enquiry into the Origins of Cultural Change*. Oxford: Blackwell.

———. 1990. "Between Space and Time: Reflections on the Geographical Imagination." *Annals of the Association of American Geographers* 80, no. 3: 418–34.

———. 1993. "From Space to Place and Back Again: Reflections on the Condition of Postmodernity." In *Mapping the Futures: Local Cultures, Global Changes,* edited by Jon Bird, Barry Curtis, Tim Putnam, George Robertson, and Lisa Tickner, 3–29. New York: Routledge.

———. 2000. *Spaces of Hope.* Berkeley and Los Angeles: Univ. of California Press.

Havel, Vaclav. 1991. *Disturbing the Peace.* New York: Vintage.

Hénaff, Marcel, and Tracy B. Strong, eds. 2001. *Public Space and Democracy.* Minneapolis: Univ. of Minnesota Press.

Herman, R. D. K. 1999. "The Aloha State: Place Names and the Anti-Conquest of Hawai'i." *Annals of the Association of American Geographers* 89, no. 1: 76–102.

Hillis, Ken. 1999. *Digital Sensations: Space, Identity, and Embodiment in Virtual Reality.* Minneapolis: Univ. of Minnesota Press.

Hollow, Michele. 2003. "Benefits of Working from Home." Available at the Discovery Communications Web site: http://tlc.discovery.com/fan sites/babystory/articles/workfromhome_39_01.html. Accessed Aug. 1, 2003.

Horkheimer, Max. [1941] 1982. "The End of Reason." *Studies in Philosophy and Social Sciences* 9: 366–88. Reprinted in *The Essential Frankfurt School Reader,* edited by Andrew Arato and Eike Gebhardt, 26–48. New York: Continuum.

Horkheimer, Max, and Theodor Adorno. [1944] 1972. *The Dialectic of Enlightenment.* Reprint. New York: Herder and Herder.

Horvath, Ronald J. 1974. "Machine Space." *Geographical Review* 64: 167–68.

Hough, Michael. 1995. *Cities and Natural Process.* New York: Routledge.

Howarth, William. 2001. "Reading the Wetlands." In *Textures of Place: Exploring Humanist Geographies,* edited by Paul C. Adams, Steven Hoelscher, and Karen E. Till, 55–83. Minneapolis: Univ. of Minnesota Press.

Jackson, Peter. 1989. *Maps of Meaning.* London: Unwin Hyman.

Janelle, Donald. 1968. "Central Place Development in a Time-Space Framework." *The Professional Geographer* 20, no. 1: 5–10.

———. 1969. "Spatial Reorganization: A Model and Concept." *Annals of the Association of American Geographers* 59, no. 2: 348–64.

———. 1973. "Measuring Human Extensibility in a Shrinking World." *Journal of Geography* 72, no. 5: 8–15.

———. 1991. "Global Interdependence and Its Consequences." In *Collapsing

Space and Time, edited by Stanley Brunn and Thomas Leinbach, 49–81. London: HarperCollins.

Janis, Irving. 1965. "The Problem of Validating Content Analysis." In *Language of Politics: Studies in Quantitative Semantics,* edited by Harold Lasswell and Nathan Leites, 55–82. Cambridge, Mass.: MIT Press.

Kar, Snehendu B., Kevin Campbell, Armando Jimenez, and Sangeeta R. Gupta. 1995–96. "Invisible Americans: An Exploration of Indo-American Quality of Life." *Amerasia Journal* 21, no. 3 (winter): 25–52.

Kellner, Douglas. 1995. "Intellectuals and the New Technologies." *Media, Culture, and Society* 17: 201–17.

———. 2004. "Habermas, the Public Sphere, and Democracy: A Critical Intervention." Available at: www.gseis.ucla.edu/faculty/kellner/papers/habermas.htm.

Knox, Paul, and John Agnew. 1989. *The Geography of the World Economy.* London: Edward Arnold.

Krishna, Anirudh. 2000. "Creating and Harnessing Social Capital." In *Social Capital: A Multifaceted Perspective,* edited by Partha Dasgupta and Ismail Serageldin, 71–93. Washington, D.C.: World Bank.

Kuhn, Thomas. 1996. *The Structure of Scientific Revolutions.* 3rd ed. Chicago: Univ. of Chicago Press.

Labov, William. 1972. *Sociolinguistic Patterns.* Philadelphia: Univ. of Pennsylvania Press.

La Haye, Yves de, ed. 1979. *Marx and Engels on the Means of Communication: The Movement of Commodities, People, Information, and Capital.* New York: International General/Bagnoletl; Paris: International Mass Media Research Center.

Lakoff, G., and M. Johnson. 1980. *Metaphors We Live By.* Chicago: Univ. of Chicago Press.

Lasch, Christopher. 1978. *The Culture of Narcissism: American Life in an Age of Diminishing Expectations.* New York: W. W. Norton.

Lévi-Strauss, Claude. 1974. *Tristes tropiques.* Translated by John Weightman and Doreen Weightman. New York: Pocket Books.

Lewis, Martin, and Kären Wigen. 1997. *The Myth of the Continents: A Critique of Metageography.* Berkeley and Los Angeles: Univ. of California Press.

Longhurst, Robyn. 2001. *Bodies: Exploring Fluid Boundaries.* New York: Routledge.

Low, Murray. 1997. "Representation Unbound: Globalization and Democracy." In *Spaces of Globalization: Reasserting the Power of the Local,* edited by Kevin Cox, 240–80. New York: Guilford Press.

Lowenthal, David. 1985. *The Past Is a Foreign Country.* Cambridge: Cambridge Univ. Press.

Lucy, John A. 1992a. *Grammatical Categories and Cognition: A Case Study of the Linguistic Relativity Hypothesis.* Cambridge, England: Cambridge Univ. Press.

———. 1992b. *Language Diversity and Thought: A Reformulation of the Linguistic Relativity Hypothesis.* Cambridge: Cambridge Univ. Press.

Lukács, Georg. [1923] 1971. *History and Class Consciousness.* Translated by R. Livingstone. Reprint. Cambridge, Mass.: MIT Press.

Lynch, Kevin. 1960. *The Image of the City.* Cambridge, Mass.: MIT Press.

Mann, Michael. 1986. *The Sources of Social Power.* New York: Cambridge Univ. Press.

Marcuse, Herbert. 1978. *The Aesthetic Dimension: Toward a Critique of Marxist Aesthetics.* Boston: Beacon Press.

Marvin, Carolyn. 1988. *When Old Technologies Were New: Thinking about Electric Communication in the Late Nineteenth Century.* New York: Oxford Univ. Press.

Marx, Karl. [1852] 1973. *The Eighteenth Brumaire of Louis Bonaparte.* In *Surveys from Exile: Political Writings,* edited by David Fernbach, 143–249. New York: Random House.

Massey, Dorren. 1994. *Space, Place, and Gender.* Cambridge: Polity.

McCarthy, Thomas. 1990. Introduction to *Moral Consciousness and Communicative Action,* by Jürgen Habermas, translated by Christian Lenhardt and Shierry Weber Nicholsen, vii–xiii. Cambridge, Mass.: MIT Press.

McDowell, Linda. 1999. *Gender, Identity, and Place: Understanding Feminist Geographies.* Minneapolis: Univ. of Minnesota Press.

McLuhan, Marshall. 1962. *The Gutenberg Galaxy: The Making of Typographic Man.* Toronto: Univ. of Toronto Press.

McLuhan, Marshall, and Quentin Fiore. 1967. *The Medium Is the Massage.* New York: Bantam.

Meyrowitz, Joshua. 1985. *No Sense of Place: The Impact of Electronic Media on Social Behavior.* New York: Oxford Univ. Press.

Mishra, Anjana Agnihotri. 2000. "Asian Indian Americans in South Florida: Values and Identity." In *Worldview Flux: Perplexed Values among Postmodern Peoples,* edited by Jim Norwine and Jonathan M. Smith, 177–97. Lanham, Md.: Lexington.

Mitchell, Don. 1995. "There's No Such Thing as Culture: Towards a Reconceptualisation of the Idea of Culture in Geography." *Transactions of the Institute of British Geographers* 19: 102–16.

———. 2000. *Cultural Geography: A Critical Introduction.* Oxford: Blackwell.

Mitchell, William J. 1995. *City of Bits.* Cambridge, Mass.: MIT Press.

Mumford, Lewis. 1961. *The City in History: Its Origins, Its Transformations, and Its Prospects.* San Diego: Harcourt Brace Jovanovich.

———. [1934] 1963. *Technics and Civilization.* With a new introduction. Reprint. New York: Harcourt, Brace and World.

Nagel, Thomas. 1986. *The View from Nowhere.* New York: Oxford Univ. Press.

Olalquiaga, Celeste. 1992. *Megalopolis: Contemporary Cultural Sensibilities.* Minneapolis: Univ. of Minnesota Press.

Oliver, Miguel de, and Michael Yoder. 2000. "Postmodernity and the Borderlands: Symbolic Democracy and Ethnic Disparities." In *Worldview Flux: Perplexed Values among Postmodern Peoples,* edited by Jim Norwine and Jonathan M. Smith, 87–109. Lanham, Md.: Lexington.

Olsson, Gunnar. 1975. *Birds in Egg.* Michigan Geographical Publication no. 15. Ann Arbor: Department of Geography, Univ. of Michigan.

———. 1988. "The Eye and the Index Finger: Bodily Means to Cultural Meaning." In *A Ground for Common Search,* edited by Reginald Golledge, Helen Couclelis, and Peter Gould, 126–37. Goleta, Calif.: Santa Barbara Geographical Press.

———. 1992. "Lines of Power." In *Writing Worlds: Discourse, Text, and Metaphor in the Representation of Landscape,* edited by Trevor Barnes and James Duncan, 86–96. New York: Routledge.

Ó Tuathail, Gearóid. 1996. *Critical Geopolitics: The Politics of Writing Global Space.* Borderlines vol. 6. Minneapolis: Univ. of Minnesota Press.

Pavlov, Ivan, and G. V. Anrep. 1927. *Conditioned Reflexes: An Investigation of the Physiological Activity of the Cerebral Cortex.* Oxford: Oxford Univ. Press.

Peet, Richard. 1977. *Radical Geography: Alternative Viewpoints on Contemporary Social Issues.* Chicago: Maaroufa Press.

Piccone, Paul. 1982. "General Introduction." In *The Essential Frankfurt School Reader,* edited by Andrew Arato and Eike Gebhardt, ix–xxi. New York: Continuum.

Poster, Mark. 2001. *What's the Matter with the Internet?* Minneapolis: Univ. of Minnesota Press.

Pred, Allan. 1979. "The Academic Past Through a Time-Geographic Looking Glass." *Annals of the Association of American Geographers* 69, no. 1: 175–85.

———. 1982. "Social Reproduction and the Time-Geography of Everyday Life." In *A Search for Common Ground,* edited by Peter Gould and Gunnar Olsson, 157–86. London: Pion.

———. 1984. "Structuration, Biography Formation, and Knowledge: Observations on Port Growth during the Late Mercantile Period." *Environment and Planning D: Society and Space* 2, no. 3: 251–75.

Rand, Ayn. 1964. "The Objectivist Ethics." In *The Virtue of Selfishness: A New Concept of Egoism,* 13–39. With additional articles by Nathaniel Branden. New York: Signet/New American Library.

Rawls, John. [1971] 1999. *A Theory of Justice.* Rev. ed. Cambridge, Mass.: Belknap Press of Harvard Univ. Press.

Relph, Edward. 1976. *Place and Placelessness.* London: Pion.

———. 1989. "Responsive Methods, Geographical Imagination, and the Study of Landscapes." In *Remaking Human Geography,* edited by A. Kobayashi and S. Mackenzie, 149–63. London: Unwin Hyman.

Rheingold, Howard. 1993. *The Virtual Community: Homesteading on the Electronic Frontier.* Reading, Mass.: Addison Wesley.

Rorty, Richard. 1998. "Justice as a Larger Loyalty." In *Cosmopolitics: Thinking and Feeling beyond the Nation,* edited by Pheng Cheah and Bruce Robbins, 45–58. Minneapolis: Univ. of Minnesota Press.

Rose, Gillian. 1993. *Feminism and Geography: The Limits of Geographical Knowledge.* Minneapolis: Univ. of Minnesota Press.

Sack, Robert David. 1980. *Conceptions of Space in Social Thought: A Geographic Perspective.* Minneapolis: Univ. of Minnesota Press.

———. 1986. *Human Territoriality.* Cambridge: Cambridge Univ. Press.

———. 1997. *Homo Geographicus: A Framework for Action, Awareness, and Moral Concern.* Baltimore: Johns Hopkins Univ. Press.

Said, Edward. 1978. *Orientalism.* New York: Pantheon.

Saussure, Ferdinand de. 1983. *Course in General Linguistics.* Edited by Charles Bally and Albert Sechehaye. Translated by Roy Harris. La Salle, Ill.: Open Court.

Saville-Troike, Muriel. 1989. *The Ethnography of Communication: An Introduction.* 2d ed. Cambridge, Mass.: Blackwell.

Schein, Richard. 1997. "The Place of Landscape: A Conceptual Framework for Interpreting an American Scene." *Annals of the Association of American Geographers* 87, no. 4: 660–80.

Schiller, Herbert. 1995. "The Global Information Highway: Project for an Ungovernable World." In *Resisting the Virtual Life: The Culture and Politics of Information,* edited by James Brook and Iain A. Boal, 17–33. San Francisco: City Lights.

Schmandt-Besserat, Denise. 1978. "The Earliest Precursor of Writing." *Scientific American* 238, no. 6: 50–59.

Sennett, Richard. 1978. *The Fall of Public Man: On the Social Psychology of Capitalism.* New York: Vintage.

———. 1980. *Authority.* New York: Random House, Vintage.

Shakespeare, William. 1975. *As You Like It.* Edited by Agnes Latham. New York: Harper and Row.

Simmel, Georg. 1961. "The Metropolis and Mental Life." In *Classic Essays on the Culture of Cities,* edited by Richard Sennett, 47–60. New York: Appleton-Century-Crofts.

Smith, Donald Eugene. 1963. *India as a Secular State.* Princeton, N.J.: Princeton Univ. Press.

Smith, Jonathan M. 1993a. "The Lie That Blinds: Destabilizing the Text of Landscape." In *Place/Culture/Representation,* edited by James Duncan and David Ley, 78–92. New York: Routledge.

———. 1993b. "Writing from the Margins: Travel Writing and the Professors." Paper presented at the Eighty-ninth Annual Meeting of the Association of American Geographers, Atlanta, Georgia, Apr.

———. 1996. "Geographical Rhetoric: Modes and Tropes of Appeal." *Annals of the Association of American Geographers* 86, no. 1: 1–20.

Smith, Neil. 1992. "Contours of a Spatialized Politics: Homeless Vehicles and the Production of Geographical Scale." *Social Text* 33: 55–81.

———. 1993. "Homeless/Global: Scaling Places." In *Mapping the Future: Local Cultures, Global Change,* edited by Jon Bird, Barry Curtis, Tim Putnam, George Robertson, and Lisa Tickner, 87–119. New York: Routledge.

———. 1996. *The New Urban Frontier: Gentrification and the Revanchist City.* New York: Routledge.

Soja, Edward. 1996. *Thirdspace: Journeys to Los Angeles and Other Real-and-Imagined Places.* Cambridge, Mass.: Blackwell.

Solnit, Rebecca. 2000. *Wanderlust: A History of Walking.* London: Viking Penguin.

Stamps, Judith. 1995. *Unthinking Modernity: Innis, McLuhan, and the Frankfurt School.* Montreal: McGill-Queens Univ. Press.

Taylor, Jonathan. 1997. "The Emerging Geographies of Virtual Worlds." *Geographical Review* 87, no. 2: 172–92.

Teather, Elizabeth Kenworthy. 1999. *Embodied Geographies: Spaces, Bodies, and Rites of Passage.* New York: Routledge.

Thrift, Nigel. 1985. "Flies and Germs: A Geography of Knowledge." In *Social Relations and Spatial Structures,* edited by D. Gregory and J. Urry, 366–403. New York: St. Martin's Press.

———. 1986. "Little Games and Big Stories: Accounting for the Practice of Personality and Politics in the 1945 General Election." In *Politics, Geography, and Social Stratification,* edited by K. Hoggart and E. Kofman, 86–143. Wolfeboro, N.H.: Croom Helm.

Thrift, Nigel, and Dean Forbes. 1983. "A Landscape with Figures: Political Geography with Human Conflict." *Political Geography Quarterly* 2, no. 3: 247–63.

Toffler, Alvin. 1980. *The Third Wave.* New York: Morrow.

Toffler, Alvin, and Heidi Toffler. 1995. *Creating a New Civilization: The Politics of the Third Wave.* Atlanta, Ga.: Turner.

Tuan, Yi-Fu. 1974. *Topophilia: A Study of Environmental Perception, Attitudes, and Values.* With a new preface by the author. New York: Columbia Univ. Press.

———. 1977. *Space and Place: The Perspective of Experience.* Minneapolis: Univ. of Minnesota Press.

———. 1978. "Literature and Geography: Implications for Geographical Research." In *Humanistic Geography: Prospects and Problems,* edited by D. Ley and M. Samuels, 194–206. Chicago: Maaroufa Press.

———. 1979. *Landscapes of Fear.* New York: Pantheon.

———. 1982. *Segmented Worlds and Self: Group Life and Individual Consciousness.* Minneapolis: Univ. of Minnesota Press.

———. 1984. *Dominance and Affection: The Making of Pets.* New Haven, Conn.: Yale Univ. Press.

———. 1986a. *The Good Life.* Madison: Univ. of Wisconsin Press.

———. 1986b. "Strangers and Strangeness." *Geographical Review* 76, no. 1: 10–19.

———. 1988. "The City as a Moral Universe." *Geographical Review* 78, no. 3: 316–24.

———. 1991. "Language and the Making of Place: A Narrative-Descriptive Approach." *Annals of the Association of American Geographers* 81, no. 4: 684–96.

————. 1996. *Cosmos and Hearth: A Cosmopolite's Viewpoint.* Minneapolis: Univ. of Minnesota Press.

————. 1998. *Escapism.* Baltimore: Johns Hopkins Univ. Press.

————. 1999. *Who Am I? An Autobiography of Emotion, Mind, and Spirit.* Madison: Univ. of Wisconsin Press.

————. 2002. *Dear Colleague: Common and Uncommon Observations.* Minneapolis: Univ. of Minnesota Press.

Turkle, Sherry. 1995. *Life on the Screen: Identity in the Age of the Internet.* New York: Simon and Schuster.

Warf, Barney, and John Grimes. 1997. "Counterhegemonic Discourses and the Internet." *Geographical Review* 87, no. 2: 259–74.

Warnke, Georgia. 1995. "Communicative Rationality and Cultural Values." In *The Cambridge Companion to Habermas,* edited by Stephen K. White, 120–42. Cambridge: Cambridge Univ. Press.

Warren, Mark E. 1995. "The Self in Discursive Democracy." In *The Cambridge Companion to Habermas,* edited by Stephen K. White, 167–200. Cambridge: Cambridge Univ. Press.

Waterstone, Marvin. 1998. "Better Safe Than Sorry, or Bettor Safe, Then Sorry?" *Annals of the Association of American Geographers* 88, no. 2: 297–300.

Webber, Melvin. 1964. "The Urban Place and the Nonplace Urban Realm." In *Explorations into Urban Structure,* edited by M. Webber, J. Dyckman, J. Foley, A. Guttenberg, W. Wheaton, and C. Wurster, 79–153. Philadelphia: Univ. of Pennsylvania Press.

Wellman, Barry, and Milena Gulia. 1999. "Virtual Communities as Communities: Net Surfers Don't Ride Alone." In *Communities in Cyberspace,* edited by Marc A. Smith and Peter Kollock, 167–94. New York: Routledge.

Wellman, Barry, Renita Wong, David Tindall, and Nancy Glazer. 1997. "A Decade of Network Change: Turnover, Mobility, and Stability." *Social Networks* 19: 27–50.

West, Cassandra. "The New Mix." Reprinted from the *Chicago Tribune* by the Black Sisters and Brothers United Network, available at: www.bsbu-network.com/Email_Archive/MIX.htm.

White, Hayden. 1996. "Storytelling: Historical and Ideological." In *Centuries' Ends, Narrative Means,* edited by Robert Newman, 58–78. Stanford, Calif.: Stanford Univ. Press.

White, Stephen K., ed. 1995. "Introduction: Reason, Modernity, and Democracy." In *The Cambridge Companion to Habermas,* edited by Stephen K. White, 3–16. Cambridge: Cambridge Univ. Press.

Whorf, Benjamin. 1956. *Language, Thought, and Reality: Selected Writings of Benjamin Lee Whorf.* Edited by John B. Carroll. Cambridge, Mass.: MIT Press.

Williams, Raymond. 1974. *Television: Technology and Cultural Form.* New York: Schocken.

Wilson, Elizabeth. 1991. *The Sphinx in the City: Urban Life, the Control of Disorder, and Women.* London: Virago.

―――. 1992. "The Invisible Flâneur." *New Left Review* 191: 90–110.

Wirth, Louis. 1938. "Urbanism as a Way of Life." *American Journal of Sociology* 44: 1–24.

Wittfogel, Karl. 1957. *Oriental Despotism: A Comparative Study of Total Power.* New Haven, Conn.: Yale Univ. Press.

Wolcott, Jennifer. 2002. "The Good Neighbor Policy." *Christian Science Monitor,* Sept. 18. Available at: www.csmonitor.com/2002/0918/p15s02-lihc.html.

Wolff, Janet. 1985. "The Invisible Flâneuse: Women and the Literature of Modernity." *Theory, Culture, and Society* 2: 37–46.

Wright, J. K. 1947. "Terrae Incognitae: The Place of Imagination in Geography." *Annals of the Association of American Geographers* 37: 1–15.

Zelinsky, Wilbur. 1973. *Cultural Geography of the United States.* Englewood Cliffs, N.J.: Prentice Hall.

Index